PROBLEM SOLVING

with | Fractions, Decimals, and Percents

Janet Pittock

 **Wright Group**
McGraw-Hill

Acknowledgments

Project Editor Barbara Bobb
Design Director Karen Stack
Design O'Connor Design
Cover Illustration Lisa Manning
Illustration Jim Dandy
Production O'Connor Design

ISBN 0-7622-1254-3

Catalog #0-7622-1254-3

Customer Service 800-624-0822

1 2 3 4 5 6 7 8   MAL   05 04 03 02 01

Contents

Overview

What are the Goals of **Problem Solving with Fractions, Decimals and Percents?**

This book provides students with a wide variety of fraction, decimal, and percent problem experiences.

Students will learn

- to use fractions, decimals, and percents interchangeably and flexibly to solve problems.

- that problems involving fractions, decimals, or percents can often be solved using problem-solving strategies that students already know.

- to choose whether to use a fraction, decimal, or percent based on the relationships within the problem and the resulting calculations.

One of the goals of all mathematics education is for students to become successful problem solvers. Toward this end, **Problem Solving with Fractions, Decimals, and Percents** places equal emphasis on content and process. Problems are designed so that in many cases, students will be able to choose from a variety of approaches. We recommend during the discussion of problems that all students' suggestions for solving be considered. This will reinforce students' understanding that various approaches can often be used to solve the same problem. Students are more likely to participate if they understand that incorrect solution paths will be valued for what can be learned from them.

This program is intended for use by fifth through eighth grade students who have studied a full range of fraction, decimal, and percent topics. The program does not teach these topics, but rather provides problem situations in which students can apply their understandings in a problem context.

Calculators can be helpful at reducing the tedium of lengthy computations. If you allow students to use calculators for problems with heavy computation loads, they may see the patterns more readily since they can focus on patterns rather than procedures.

 Problems that are particularly appropriate for calculator use are marked with this calculator illustration on the teaching notes pages, but many of the problems in this book are best done without a calculator to allow students to focus on relationships and building number sense. If students use calculators, the impetus to think about relationships is greatly decreased. Students may miss out on building their sense of the relationships between fractions, decimals, and percents, a valuable skill for the many situations when students need a sense of the answer without going through the exact calculations.

How is **Problem Solving with Fractions, Decimals, and Percents** organized?

Sections

Each section has a different mathematical focus. Section 1 focuses on naming parts of a whole or set with a fraction, decimal, or percent. Section 2 provides problems that require students to add, subtract, multiply, and divide with fractions. Section 3 is similar to Section 2 with the emphasis on decimal operations. Section 4 focuses on percents and operations with percents. In Section 5, students solve problems that use all three representations. They focus on choosing a representation that is meaningful and that simplifies the computations. Section 6 focuses on ratios in problem settings, where students use fractions, decimals, and percents to solve problems involving rate, unit rate, and simple probabilities.

Each section includes integrated information about instruction, applications, and assessment. Within each section there are three levels of application activities. In order of increasing difficulty, they are *Try It Out, Stretch Your Thinking,* and *Challenge Your Mind.*

Instruction

Each section begins with a two-page teacher *Introduction.* There is a discussion of the types of problems in the section and the mathematics that students will encounter. These are followed by problems utilizing various problem-solving strategies and solution paths similar to those students will use in the section. Each introductory section also contains five *Thinking About...* problems that focus students' attention on number sense and estimation.

The first problem in each section, *Solving Problems with…*, focuses students' attention on real-world problems that can be solved using that particular skill. Throughout the program, a variety of strategies and solution paths are presented in these problems.

Applications

The level of difficulty increases from one set to the next. *Try It Out* presents problems in which basic understandings of the section concept are presented. *Stretch Your Thinking* invites students to look beyond basic applications to more sophisticated use of the concept. *Challenge Your Mind* presents fewer, more complex problems than the previous two applications. *Wrap It Up* is an application in which students either apply what they have learned throughout the section or collect data to solve a problem that draws on skills that are the focus of the section. *Wrap It Up* activities are designed to be accessible to all students.

Assessment

There are multiple opportunities for assessment throughout **Problem Solving with Fractions, Decimals and Percents**. Each *Try It Out, Stretch Your Thinking,* and *Challenge Your Mind* page contains a suggestion for informal assessment. These are in the form of questions to pose to students in order to focus their attention on one important aspect of the problem-solving process.

Each *Wrap It Up* contains an assessment rubric that sets a performance standard for that activity. You may wish to review the rubric with students before they complete the *Wrap It Up*. This will let them know what is required to demonstrate a satisfactory level of performance. *Wrap It Up* problems can become part of students' math portfolios.

Using the Student Pages

Many of the problems can be approached by students working either independently, with a partner, or in a cooperative group. The *Solving Problems with…* problem is designed to be a teaching problem, presented as a teacher-led activity. You may also find that selected problems or sections are particularly appropriate at a given time, such as when a topic relates well to your daily math lessons.

- **Problem of the Day** The *Thinking About* problems found in the Introduction to each section can be used to check whether students possess the prerequisite skills needed for the problems in that particular section. Have students solve one problem per day as a warm-up.

- **Problem of the Week** One or two of the more challenging problems from *Stretch Your Thinking* or *Challenge Your Mind* can be presented as a problem of the week. You might post the problem on a class bulletin board. Solutions can be discussed as a whole-class activity and perhaps posted on the bulletin board with the original problem.

- **Partners or Cooperative Groups** Students can work one or more problems in a section, arrive at solutions that group members agree upon, and present their solutions to the rest of the class.

- **Whole Class Lesson** As you look through this book you may see problems and skills that fit into your regular lessons or that integrate or enhance a particular curricular theme or unit. You may wish to use a selected page as your math lesson for the day.

Using the Teaching Pages

The main teaching technique in **Problem Solving with Fractions, Decimals, and Percents** is the use of thought-provoking questions asked at the appropriate time. Much of the benefit to be gained from the program comes from students recognizing good questions. Over time, this will lead them to begin to ask themselves the kinds of questions needed to analyze a problem. However, this is a lengthy process and you should not look for results in the short run. A commitment to using the program and its approach to problem solving will lead to success in the long run.

Problem-Solving Techniques

The Ten Solution Strategies

1. **Act Out or Use Object** Acting out a problem helps you see the data and watch the solution process. You can also use objects to act out a problem.

2. **Make a Picture or Diagram** Drawing a picture or a diagram can help you see and understand the data in a problem.

3. **Use or Make a Table** Tables help you keep track of data and see patterns.

4. **Make an Organized List** Making a list makes it easier to review the data and helps you organize your thinking.

5. **Guess and Check** Guess the answer and check to see if it is correct. If your guess is incorrect, use what you learned to make more reasonable guesses until you solve the problem.

6. **Use or Look for a Pattern** Recognizing patterns can help you predict what will come next.

7. **Work Backward** Sometimes you can start with the data at the end of the problem and work backward to find the answer.

8. **Use Logical Reasoning** Use logical reasoning to solve problems that have conditions such as "if something is true, then …." Making a table often helps to solve logical reasoning problems.

9. **Make it Simpler** When a problem looks very hard, you can replace large numbers with smaller ones or reduce the number of items and try to solve that problem first.

10. **Brainstorm** When you cannot think of a strategy to use or a way to solve a problem, it's time to brainstorm. Stretch your thinking and be creative.

The Four-Step Plan

Step 1 Find Out	Find out what question the problem is asking you to answer, what information you have, and what information you need to get.
Step 2 Make a Plan	Make a plan that will help you solve the problem. Often, choosing a strategy is the first part of making a plan.
Step 3 Solve It	Solve it by using your plan to work through the problem until you find the answer to the question. If your plan does not work, try another approach.
Step 4 Look Back	Look back at the problem. Check that your solution answers the question and is a reasonable answer.

Problem-Solving Recording Sheet

Page: _____   Problem Number: _____

Find Out	○ What question must you answer to solve the problem?
	● What information do you have?
Make a Plan	○ How will you use the information you have to solve the problem?
Solve It	● Complete the solution, using your plan. Record your work.

The Solution: _____

Look Back	○ Is your answer reasonable?
	● How can you check your solution?

The focus of this section is on the different ways of representing parts of a whole using fractions, decimals, and percents. Students will deepen their understanding of the part/whole relationship as they learn to work flexibly with the various representations of rational numbers.

Understanding Using Representations

As students investigate the use of fractions, decimals, and percents to represent part/whole relationships, they need to develop facility converting from one representation to another. By having a wide variety of experiences in which to use different representations of rational numbers, students begin to develop a sense about the advantages and disadvantages of the various representations of a quantity in different situations.

Students will find and compare parts of regions and express the areas as fractions, decimals and percents. They will also locate and place points between 0 and 1, or some percent of the way between two numbers on a number line. These models help students view fractions, decimals, and percents as different ways to represent mathematical relationships.

Give a fraction, decimal, and percent name to each part of the rectangle.

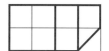

To solve this problem, students must:

- know the area or number of units in the whole and each part.

- be able to represent the parts and whole as a fraction, decimal, and percent.

   The Rosario family has many animals. They have 15 chickens, 4 dogs, 6 cats, 2 cows, and 24 fish. What fraction of their mammals are cats? What percent of the animals are chickens?

In a problem such as this, students must:

- be clear about what is the whole. In one case, the whole is 12 mammals, and in the other, the whole is 51 animals.

- be able to find the percent when they know the part and whole.

   What number, expressed as a decimal, corresponds to point *A*? Place point *B* 50% of the way between 2 and 5. What decimal number corresponds to *B*? Place point *C* two-thirds of the way between 2 and 5. What decimal number corresponds to it?

In a problem such as this, students must:

- be able to name decimal numbers between whole numbers.

- realize that to name something 50% of the way between two numbers means that they must find the distance between the two, find 50% of that, and add (or subtract) the 50% distance to the appropriate place on the number line. A similar kind of thinking is applied to the fractional measure.

As students work with fraction, decimal, and percent representations, ask these key questions, as needed:

- What is the whole? What is the part?

- What information do you have? What do you need? If you have missing information, how can you get it?

- Does the problem have multiple steps?

- Is there more than one answer?

Using Fraction, Decimal, and Percent Representations in Problems

Students should be encouraged to use some of their own strategies when comparing or ordering rational numbers. For example, some students may find it easier to convert fractions to decimals, whereas others may use benchmark fractions or even visual images. If asked which is larger $\frac{2}{3}$ or $\frac{3}{4}$, one student may say that $\frac{3}{4}$ is larger because each fraction is one piece less than a whole but the thirds are larger than fourths. Another student may think 0.75 is greater than $0.6\overline{6}$.

Solving a Problem by *Making a Picture or Diagram*

Students can use diagrams to determine mathematical relationships. In this example a 10 × 10 grid helps students see the connection between decimal, fraction and percent representations.

Example: Shade regions of a 10 × 10 grid to represent 10%, 0.25, and $\frac{1}{3}$.

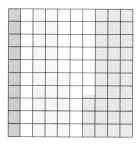

To solve this problem, students will need to realize that the whole is made up of one hundred grid squares. They will have to decide how many squares to shade to represent 0.25, $\frac{1}{3}$, and 10%. In each case, they will find the part of 100 that matches each representation.

Assessment

☑ **Informal Assessment** A suggestion for informal assessment will be found on each *Try It Out, Stretch Your Thinking,* and *Challenge Your Mind* page. The recommended question will help focus students' attention on one part of the problem-solving process.

Assessment Rubric An assessment rubric is provided for each *Wrap It Up.* Students' completed work may be added to their math portfolios.

Thinking About Representations

These problems ask students to use fraction, decimal, and percent representations of mathematical situations. Present one problem a day as a warm-up. You may choose to read the daily problem aloud, write it on the board, or create a transparency.

1. Draw a circle on the board. Shade $\frac{1}{2}$. Ask: **What part of the circle is shaded? How many different ways can you write the number that represents the shaded portion? List some of them.**
   ($\frac{1}{2}$ and fractions equivalent to $\frac{1}{2}$, 0.5, 50%)

2. Draw 24 **X's** on the board, then circle a set of 6 of them. Ask: **What fractional part of the whole is not circled? Write that fraction in different ways. Be sure to use decimals and percents.**
   ($\frac{3}{4}, \frac{6}{8}, \frac{9}{12}$, 0.75, 75%; If students have difficulty, draw their attention to 24 as the whole set and discuss dividing it into four groups of 6.)

3. Draw or place a transparency of a 10 × 10 grid on the overhead. Shade 20 squares. Ask: **What fractional part of the whole is shaded? Write that fraction in as many different ways as you can.**
   ($\frac{20}{100}$ and representations equivalent to $\frac{20}{100}$ such as 20%, $\frac{2}{10}, \frac{1}{5}$, 0.20) For a challenge, shade in twelve and a half squares and ask students the same question.
   ($\frac{25}{200}, \frac{1}{8}$, 0.125, 12.5%, $12\frac{1}{2}$%.)

4. Write $\frac{1}{2}$ and $\frac{3}{4}$ on board. Say: **Write as many numbers as you can in the next 3 minutes that are between these two numbers. When you are finished, rewrite each one in a different form.**
   (Answers will vary. One example is $\frac{5}{8}$. In a different form it is 0.625 or 62.5%.)

5. Write $\frac{2}{3}$, 80%, 0.25, $33\frac{1}{3}$%, and $\frac{1}{5}$ on the board. **Put these numbers in order from largest to smallest. Justify your answer.**
   ($\frac{1}{5}$, 0.25, $33\frac{1}{3}$%, $\frac{2}{3}$, 80%; Students will have many different justifications. Accept all that are mathematically correct.)

1 USING REPRESENTATIONS

Solving Problems with Fractions, Decimals, and Percents

This lesson will help students focus on naming the same quantity with three different representations—a fraction, a decimal, and a percent.

Using the Four-Step Method

Find Out	○ The problem asks students to identify the fractional portion for each color and then represent each fraction as a decimal and a percent.
	● Some students may see that dividing the gray section into thirds and the white section in half gives five equal sized pieces. The black will need to be divided into five equal pieces also. Others may have different ways to divide the square into equal portions.
	○ Students should easily see that $\frac{1}{2}$ of the square is black, a little more than $\frac{1}{4}$ is gray and a little less than $\frac{1}{4}$ is white. Other estimates are acceptable. Students may express their estimates as fractions, decimals, or percents.
Make a Plan	● Students know the total number of equal portions they have divided the square into and the number of portions of each color. They need to express each color as a fraction, decimal, and percent of the whole.
	○ Students will compare the number of portions in each area to the total number of portions. They will likely begin with the representation with which they feel most comfortable— fraction, decimal or percent—and then convert to the other representations.
Solve It	● The black part is $\frac{1}{2}$, 50%, and 0.5. The gray part is $\frac{3}{10}$, 30%, and 0.3. The white part is $\frac{1}{5}$, 20%, and 0.2. Some students will begin with fractions having denominators of 10, simplify them, and then convert to percents and decimals. Others will realize that the tenths can be more easily converted to percents and decimals before simplifying the fractions.
Look Back	○ Have students compare their results to their estimates. Ask which form of representation is easiest to use when making comparisons. If they know that $\frac{1}{4}$ is 25% (the number used in their estimate), they may find that using the percents is the quickest way to compare.
	● Students can check their solutions by calculating the actual areas of the square and the pieces to see if the fractional relationships are the same as their answers.

1 USING REPRESENTATIONS

Solving Problems with Fractions, Decimals, and Percents

The students in Mr. Cruz's class are making a quilt for the local homeless shelter. Each student is making a quilt square like the one shown, but larger. The PTA has agreed to purchase the fabric. What fractional portion of the fabric will be black? Gray? White? Write your answers as decimals and percents.

Find Out	○ What does the problem ask you to do?
	● How can the square be divided so that you can figure out fractional parts for each color?
	○ What are reasonable estimates?
Make a Plan	● What information do you have? What information do you need?
	○ How will you use the information to find the answer?
Solve It	● Find the answer. Use your plan. Keep a record of your work.
Look Back	○ How does your answer compare with your estimate?
	● How could you check if your answer is right?

1 | USING REPRESENTATIONS
Try It Out

1. This problem requires students to use a whole that is greater than one. They also have to find the value of the whole, given the size and value of a part.

Make a Plan To help students plan their solution, ask:

- What does the problem ask you to find? (How many girls attend the school.)

- What information do you have? How can you use that information to find the number of girls? (There are 720 boys represented by 45 squares. 55 squares represent girls. Figure out how many students are represented by one square and use that number to calculate how many girls.)

- What process will you use to find out how much each square is worth? (Students know that 45 squares are worth 720. One square is worth 720 ÷ 45.)

Solution Path

880 girls

The number of students represented by one square is 720 ÷ 45 = 16. Since 55 squares represent girls, there are 16 × 55 or 880 girls. Some students may calculate the total number of students and subtract the number of boys to find the number of girls. 16 × 100 = 1,600 and 1,600 − 720 = 880.

2.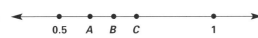

A is at 0.6, **B** is at $\frac{2}{3}$ or 0.67, **C** is at 0.75.

Students' explanations will vary. They might note that **B** is to the right of **A**, because 0.67 is greater than 0.6. **C** at 75% is halfway between 0.5 and 1.0.

3. 50%, 3.75, $\frac{1}{3}$

D is at 5, so 50% of the 0 to 10 strip is to the left of **D**. E is more than halfway between 3 and 4, so it might be about 3.75. **F** is more than halfway between 6 and 7. If it is at 6.67, that is $\frac{2}{3}$ of the strip, so $\frac{1}{3}$ of the strip is to its right. Since students must estimate the actual corresponding numbers for **E** and **F**, accept answers that are close.

☑ **Informal Assessment** Ask: Where on the 0 to 10 strip would you place 40% of 10? Why? (At the end of the fourth block, because it is $\frac{4}{10}$ of the way between 0 and 10 and $\frac{4}{10}$ is 40%.)

---

4. $\frac{21}{36}$ or $\frac{7}{12}$, 58.3%

The key here is "What is the whole?" In this case, the whole is all the cakes, which have a total of 36 pieces cut and 21 pieces eaten.

5. $\frac{5}{8}$ took a bus, 4.2% flew, $\frac{7}{8}$ or 87.5% took transportation other than a car

$\frac{30}{48}$ is $\frac{5}{8}$, 2 ÷ 48 = 0.0416, $\frac{(48-6)}{48}$ is $\frac{42}{48}$ or 0.875.

Problem Solving with Fractions, Decimals, and Percents

1 | USING REPRESENTATIONS

Try It Out

1. The large square represents the total number of students in Amy Tan Middle School. The shaded squares represent the 720 boys. How many girls attend Amy Tan Middle School?

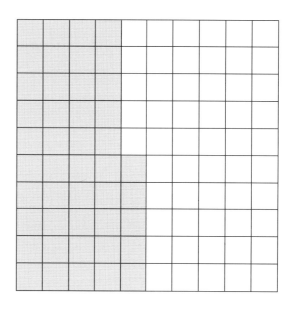

2. Estimate the number that corresponds to point **A**. Insert point **B** to correspond to approximately $\frac{2}{3}$. Insert point **C** to correspond to 75% of 1. Justify your answers.

0.5 **A**   1

3. What percent of the strip is to the left of **D**? Estimate the decimal that corresponds to **E**. Approximately what fraction of the strip is to the right of **F**?

0   **E**   **D**   **F**   10

4. Three 10-inch layer cakes had been cut into 12 pieces each for the teacher-appreciation luncheon. After lunch there were 3, 8, and 4 pieces left uneaten from the three cakes, respectively. What fraction of the cakes was eaten? What percent?

5. Of 48 ballplayers who went to the championships, 30 took a bus, 6 carpooled, 2 flew, and the rest went by train. What fraction of the players took a bus? What percent flew? What fraction took transportation other than a car? What percent?

1 | USING REPRESENTATIONS

Stretch Your Thinking

1. Make a Plan To help students understand the problem, ask:

- How long is the side of one small square in the diagram? (30 feet)

- What does the shaded portion represent? (the new school)

- How can you find the area of the rectangular lot? (Answers will vary. One answer is to multiply the length by the width. Another is to find the area of one square of the grid and multiply that by the total number of grid squares.)

Solution Paths

There will be 82.9% land left for the sports' fields and parking lots. This meets the board's requirements.

- Students can find the actual area of the school and the land and then use *part divided by the whole* to find the percent. Using *length × width = area*, the school has an area of 120 ft × 90 ft or 10,800 square feet. The total land area represented by the larger rectangle has an area of 420 ft × 150 ft or 63,000 square feet. The school area is 10,800 ÷ 63,000 = 0.171 or 17.1% of the total. That leaves 100% – 17.1% = 82.9% of the land for sports fields and parking.

- Students may also count the unshaded grid squares and compare that to the total number of grid squares to find the percent.

There are 58 grid squares not shaded out of a total of 70. 58 ÷ 70 = 0.829 or 82.9%

---

2. 750

The 116 shaded squares represent 116% which is 870 fish. So, each square represents 870 ÷ 116 or 7.5 fish. Last year is represented by 100 squares, so 7.5 × 100 is 750 fish.

☑ **Informal Assessment** Ask: How many squares would be shaded if this year there were 600 fish instead of 870 fish? (80 squares because 600 is $\frac{4}{5}$ or 80% of 750. Students might also figure out how many grid squares of 7.5 make up 600 by dividing 600 ÷ 7.5 = 80.)

3. a. Answers will vary.
One possible way:

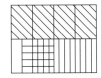

b. Answers will vary.
One possible answer is $\frac{1}{8} + \frac{1}{16} = \frac{3}{16}$.

c. Answers will vary. One possible answer is
$\frac{1}{8} + \frac{1}{8} + \frac{1}{8} = 0.375$.

d. Answers will vary. One possible answer is
$\frac{1}{8} + \frac{1}{8} + \frac{1}{8} + \frac{1}{8} = 50\%$.

---

4. $\frac{1}{12}$ One possible answer:

The remaining part is half of $\frac{1}{6}$ or $\frac{1}{12}$. Some students might see it as $\frac{1}{3}$ of $\frac{1}{4}$ which is also $\frac{1}{12}$.

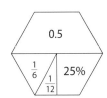

5. 7, 22%

The least fraction with a value between 0.15 and 0.36 is $\frac{5}{33}$ with a value of $0.15\overline{15}$. The greatest is $\frac{11}{33}$ with a value of $0.\overline{33}$. There are 32 fractions in all that have denominator 33 between zero and 1. 7 ÷ 32 = 0.219 or 22%.

---

1 | USING REPRESENTATIONS
Stretch Your Thinking

1. The town of Lenape plans to build a new school on land that is 150 ft by 420 ft. The planned school is 90 ft by 120 ft. The school board wants at least 80% of the land for sports fields and parking lots. If they use this school design, what percent of the land will be left?

Is that enough for the school board's requirements?

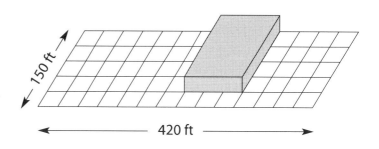

2.

The shaded squares represent the number of fish in the pond this year. It is an increase of 16% over last year. If there are 870 fish this year, how many fish were there last year?

3. Fold a piece of paper in fourths and then in half. Unfold it and fold one of the eight rectangular spaces in half.

a. Using your folds as guidelines, shade and mark $\frac{3}{16}$, 0.375, and 50%. Be creative in the way you mark the paper because some of your parts may have to overlap.

b. Write a fraction equation for the parts that make up your $\frac{3}{16}$.

c. Write a decimal equation for the parts that make up your 0.375.

d. Write a percent equation for the parts that make up your 50%.

4. Divide a regular hexagon into parts representing 25%, 0.5, and $\frac{1}{6}$. Mark the parts with the correct name given. Name the remaining part. Explain your thinking.

5. How many fractions with a denominator of 33 have values between 0.15 and 0.36? What percent of fractions between zero and one with denominators of 33 does this represent?

1 | USING REPRESENTATIONS
Challenge Your Mind

1. Find Out To help students understand the problem, ask:

- How many faces, edges, and vertices does the cube have? (6, 12, 8)

- Where are the unit cubes that have 3 painted faces positioned in the original cube? (the corners)

- How can you use the information about faces, edges, and vertices to help you solve the problem? (In combination with the original length of the edges, the number of faces, edges, and vertices, are mathematically related to the number of unit cubes with different numbers of faces painted.)

Solution Paths

12.5% of the unit cubes have 3 faces with paint, 37.5% have 2 faces with paint, 37.5% have 1 face with paint, and 12.5% have no faces with paint; The 8 cubes containing the vertices of the original cube have 3 faces of paint and 8 ÷ 64 is 12.5%. The two cubes on each edge of the original cube each have 2 faces with paint. Since there are 12 edges, that makes 12 × 2 or 24 unit cubes with 2 faces of paint. 24 ÷ 64 is 37.5%. The 4 cubes on each face that are not on the perimeter of the face each have 1 face painted. Since there are 6 faces, that is 6 × 4 or 24 unit cubes with 1 face painted. 24 ÷ 64 is 37.5%. And the rest of the cubes, 64 − 8 − 24 − 24 = 8 so 8 have no faces painted. 8 ÷ 64 is 12.5%.

☑ **Informal Assessment** Ask: Suppose you began with a 2 × 2 cube. What percent of the 1 × 1 unit cubes would have 3 painted faces? (100% because all 8 cubes would be vertex cubes.)

---

2. a. 17%; The area of the overlap is 1 × 1.2 or 1.2. The area of the figure is the area of the 2.3 × 1.5 rectangle, plus the area of the 2 × 2.4 rectangle minus the area of the overlap since it was counted twice. That is 3.45 + 4.8 − 1.2 or 7.05. 1.2 ÷ 7.05 is 0.17.

b. From part **a**, students know that the area of the whole figure is the sum of the two rectangles minus the area of overlap or 8.25 − **A**. For this part, students have to find the case where **A** ÷ (8.25 − **A**) = 25%. Students may **Guess and Check** and **Use a Table** to keep track of the percentages for various values of **A**. They will find that when **A** is 1.65, the overlap is 25% of the whole area (1.65 ÷ [8.25 − 1.65] = 0.25). There are several ways to position the two rectangles to get an overlap with an area of 1.65. The figure shown here has an overlap with dimensions 1.65 × 1. Other dimensions that work include 2 × 0.825, 1.5 × 1.1, and more.

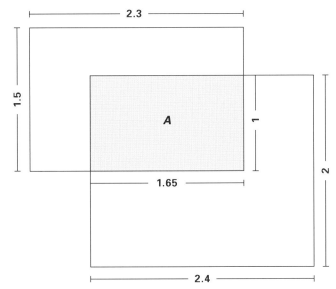

1 | USING REPRESENTATIONS

Challenge Your Mind

1. Suppose you have a 4 × 4 × 4 cube, painted on all six faces. Then you cut the large cube into 1 × 1 × 1 unit cubes. Now you have 64 cubes. Some cubes have three faces with paint on them, some have two faces with paint, some have one face with paint, and some have no faces with paint.

What percent of the cubes have 3 faces painted? 2 faces? 1 face? 0 faces?

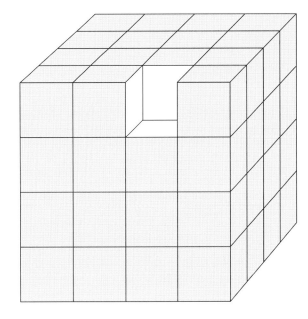

---

2. a. What percent of the figure is shaded?

   b. If you use the same two rectangles, how could you position them so that the shaded overlap is 25% of the area? Draw the new figure, and tell why you think that the shaded overlap is 25% of the area.

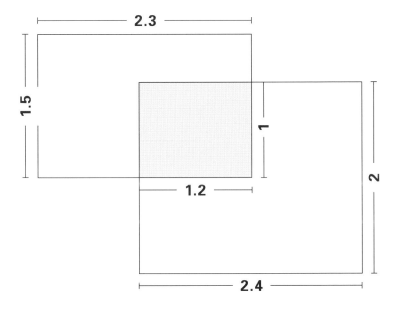

1 | USING REPRESENTATIONS
Wrap It Up

A Fraction by Any Other Name...

Discuss the problem with the class before they begin to work on their own. Ask:

- What does the problem ask you to do? (Divide the 6 × 4 rectangle into 6 pieces. All of the pieces must be different sizes. Then name each piece with a fraction, decimal, and percent.)

- How can you use the dots to help you figure out the size of your pieces? (The dots are corners of squares. You can figure out the number of squares in each piece and compare that to the 24 squares that make up the whole rectangle.)

- How can you tell the area of a triangular piece? (Some triangles are half of a rectangle. Identify the rectangle that the triangle halves. Other triangles may be part of two rectangles. Identify both rectangles, calculate half of each and add together.)

- How can you tell the area of a more complicated shape? (More complicated figures may have to be divided into several triangles and squares. Each of the separate areas is then added to find the area of the nonregular shape. There are some shapes that can not be calculated by adding areas together. In these cases suggest that students try subtracting all of the areas outside of the one that is under consideration from 24.)

Solutions
Answers will vary.

Possible Solution Paths
Students will find the area of each piece and write a fraction that represents the part that piece is of the whole. Some students will divide the

rectangle into smaller pieces so that their fractions will have denominators of 48 or even 96. Others may write a fraction within a fraction. For example, if the area of one of the pieces is $3\frac{1}{2}$, the student may write $3\frac{1}{2}$ over 24. Suggest that students multiply both numerator and denominator by the same number. The student could choose to multiply each by two to get $\frac{7}{48}$.

Assessment Rubric

3 The student accurately partitions the rectangle into at least six different pieces not all of which are rectangular or have all right angles; names each piece with the correct fraction in simplest form, decimal and percent; and justifies the fractional parts clearly and correctly.

2 The student partitions the rectangle into six pieces, but two of them may be the same or they may all be very simple shapes; names each fraction piece with the correct fraction (although it may not be in simplest form), decimal, and percent; and justifies most of the names.

1 The student partitions the rectangle into pieces, some of which may be the same; names the fraction pieces, but may make mathematical errors in calculating the areas or may be missing some of the fraction, decimal, or percent names; and justifies few or none of the names.

0 The student may partition the rectangle into pieces, but is unable to accurately name the fraction that each piece represents; and therefore also unable to justify the fractional names.

1 | USING REPRESENTATIONS

Wrap It Up

A Fraction by
Any Other Name...

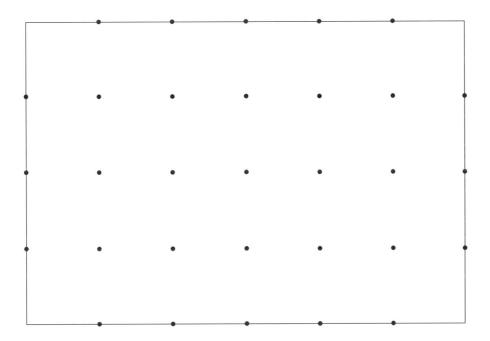

Use the rectangle shown. Partition the rectangle into at least 6 pieces, with no two pieces the same size. Be creative. Try to draw some shapes that are not just made up of full squares, but be sure you can prove the area of each shape.

Label each piece with a letter. Write the letters and the fractional part each piece is of the total rectangle. Write the fractions in simplest terms. Also, write the equivalent decimal and percent for each of the fractions.

How did you figure out the fractional parts? Explain your thinking for each of your six pieces.

2 | USING FRACTIONS
Teaching Notes

This section focuses on the solving problems using fractions. Students will compare part to whole, and will add, subtract, multiply, and divide with fractions.

Understanding Using Fractions

As students prepare to use fractions in problem-solving situations, they need to be comfortable expressing mathematical relationships with fractions.

Students will need to know not only how to add, subtract, multiply, and divide fractions, but have an understanding of what those operations mean when performed with fractions. Throughout this section, students express fractions in simplest terms.

Consider the following problems:

Cho has a 4 ft $\times$ 6 ft garden plot. She wants to plant squash in $\frac{1}{2}$ of it, peas in $\frac{1}{3}$ of it, and tomatoes in the rest of it. What fraction of the garden will be tomatoes? Draw one possible way that Cho could organize her garden.
To solve this problem, students will need to:

- find the difference between 1 and the sum of the two known fractions.

- be able to represent a 4 ft $\times$ 6 ft garden plot, and divide it accurately into $\frac{1}{2}, \frac{1}{3}$, and $\frac{1}{6}$.

Akeem and his club are making bookmarks out of ribbon for a fundraiser. Each bookmark requires $9\frac{3}{4}$ in. of ribbon. They have three 5-yd rolls of donated ribbon. How many bookmarks can they make?
In a problem such as this, students must:

- convert inches to yards or yards to inches.

- divide by a fraction.

- realize that they do not have 15 continuous yards of ribbon, but three 5–yard lengths of ribbon.

- evaluate the meaning of the remainder.

When Julio graduated from college, he was half his mother's age. At that time, Julio's son, Marco, was $\frac{1}{6}$ Julio's age and $\frac{1}{12}$ his grandmother's age. Julio was 25 when his son was born. How old were Julio, Marco, and Julio's mother when Julio graduated from college?
In a problem such as this, students must:

- realize that while the fractional relationship of the ages changes as the years progress, the number of years between the ages does not. Marco will always be 25 years younger than Julio.

- list fractions equivalent to $\frac{1}{6}$ to find the fraction whose denominator is 25 more than the numerator.

- multiply or use ratios to find Julio's mother's age given that when Julio is 30, he is half his mother's age.

As students work with fractions, ask these key questions, as needed:

- What is the numerator? Denominator? What do they represent in this problem?

- Do you add, subtract, multiply, or divide?

- Does the problem have multiple steps?

- Is there more than one answer?

Solving Fraction Problems

Your students will be more successful if they can apply logical reasoning and combine general problem-solving strategies with methods specific to solving problems with fractions. Refer to *page vi* for a discussion of strategies.

Solving a Problem by *Making an Organized List*

An organized list is particularly helpful in problems that use ratios.

Example: Jasmine is half her Aunt Tasha's age. Tasha is $\frac{2}{3}$ Jasmine's mother's age. Tasha was 12 when Jasmine was born. How old are Tasha, Jasmine, and Jasmine's mother?

Tasha is 12 years older than Jasmine, and now Jasmine is half Tasha's age. Begin by making an organized list of fractions equivalent to $\frac{1}{2}$: $\frac{1}{2}, \frac{2}{4}, \frac{3}{6}, \frac{4}{8}, \frac{5}{10}, \frac{6}{12}, \frac{7}{14}, \frac{8}{16}, \frac{9}{18}, \frac{10}{20}, \frac{11}{22}, \frac{12}{24}, \frac{13}{26}, \frac{14}{28}, \frac{15}{30}$. Identify the fraction in which the denominator is 12 more than the numerator—$\frac{12}{24}$. Therefore, Jasmine is 12 and Tasha is 24. Tasha is $\frac{2}{3}$ Jasmine's mother's age. The fraction equivalent to $\frac{2}{3}$ with a numerator of 24 is the fraction $\frac{24}{36}$. Therefore, Jasmine's mother is 36.

Jasmine is 12, Tasha is 24, and Jasmine's mother is 36.

Solving a Problem by *Making a Picture or Diagram*

Using a diagram can help students make sense of the information in a problem.

Example: Steve, Jorge, and Kwan ran laps in PE. Jorge ran $1\frac{1}{2}$ mi before he walked. Steve ran $\frac{2}{3}$ the distance Jorge did. Steve ran $\frac{4}{9}$ Kwan's distance. How far did Steve and Kwan run?

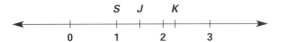

Making a diagram to show the approximate relationships between the distances helps students estimate the answers.

Steve ran $\frac{2}{3}$ of $1\frac{1}{2}$ mi. Since $1\frac{1}{2}$ mi $= \frac{3}{2}$ then $\frac{2}{3}$ of $\frac{3}{2}$, is $\frac{2}{2}$ mi or 1 mi. If 1 mi is $\frac{4}{9}$ of Kwan's distance then $\frac{1}{9}$ is $\frac{1}{4}$ of a mile so $\frac{9}{9}$ would be $9 \times \frac{1}{4}$ or $\frac{9}{4}$ or $2\frac{1}{4}$ mi.

Steve ran 1 mi. Kwan ran $2\frac{1}{4}$ mi.

Assessment

☑ **Informal Assessment** A suggestion for informal assessment will be found on each *Try It Out, Stretch Your Thinking,* and *Challenge Your Mind* page. The recommended question will help focus students' attention on one part of the problem-solving process.

Thinking About Fractions

These problems require students to use number sense and estimation skills as they prepare to use fractions to solve problems. Present one problem a day as a warm-up. You may choose to read the daily problem aloud, write it on the board, or create a transparency.

1. **Jamil cut the casserole into equal pieces. He ate 2 pieces, his brother ate 3 and his sister ate 1. Their mother found half of the casserole gone. Into how many equal pieces did Jamil cut the casserole? What fractional part is each piece?**
   ($12, \frac{1}{12}$; $2 + 3 + 1$ pieces were eaten. Those 6 pieces are $\frac{1}{2}$ of the casserole, so the whole casserole was cut into 2×6 or 12 pieces. Since the pieces were equal, each one is $\frac{1}{12}$ of the whole.)

2. 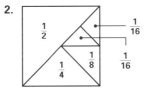 Draw the square and its parts, but not the labels. Say: **State a fraction in simplest form for each of the parts.** Direct students to choose any two parts of the square and ask: **What fractional part of the square do these two parts represent?**

   (See labels for answer to first part, $\frac{1}{2} + \frac{1}{4} = \frac{3}{4}, \frac{1}{2} + \frac{1}{8} = \frac{5}{8}, \frac{1}{2} + \frac{1}{16} = \frac{9}{16}, \frac{1}{4} + \frac{1}{8} = \frac{3}{8}, \frac{1}{4} + \frac{1}{16} = \frac{5}{16}, \frac{1}{8} + \frac{1}{16} = \frac{3}{16}, \frac{1}{16} + \frac{1}{16} = \frac{1}{8}$.)

3. Draw three rectangles on the board. Say: **Use the rectangles to show and explain the meaning of $3 \div \frac{1}{2}$.** (See how many $\frac{1}{2}$s fit into three rectangles or how many times you can subtract $\frac{1}{2}$ from 3; 6.)

4. Draw eight dots. Ask: **What is $\frac{1}{4}$ of this group? What mathematical operation could you use to find that answer?** Next ask: **What is $\frac{3}{4}$ of 40? What are two ways to find that answer?**
   (2, multiplication, 30, draw 4 groups of 10 dots and circle three of them, $\frac{3}{4} \times 40 = 30$; To find $\frac{1}{4}$ of 8 you can find $\frac{1}{4} \times 8 = 2$.)

5. **Is $\frac{2}{5}$ of 98 more than 50 or less than 50?**
   (Less; $\frac{2}{5}$ is less than $\frac{1}{2}$, and 50 is more than $\frac{1}{2}$ of 98. Another explanation is that $\frac{2}{5}$ of 100 is 40, so $\frac{2}{5}$ of 98 must be less than 40, therefore less than 50.)

2 | USING FRACTIONS

Solving Problems Using Fractions

This lesson involves adding and dividing fractions. Emphasize that students may solve this in several different ways. Allow your students to share mathematically correct methods and discuss which ones are efficient. Emphasize that there is no one best method, that the method that makes sense to each student is the best method for that student.

Using the Four-Step Method

Find Out	○	Students need to find out how many 8-ft pieces of stock it takes to have 400 pieces of wood each $13\frac{5}{8}$ in. long.
	●	One estimate is more than 50 pieces of stock; Since $13\frac{5}{8}$ is more than a foot, fewer than 8 kit pieces can be made from each 8-ft piece of stock. So it will take more than $400 \div 8$ or 50 pieces of stock. Accept all estimates that can be mathematically justified.
Make a Plan	○	Students know that it takes $13\frac{5}{8}$ in. plus $\frac{1}{16}$ in. for the cut or $13\frac{11}{16}$ in. for each base. Since the stock comes in 8-ft lengths, they need to know how many pieces can be cut from each 8-ft length and then how many 8-ft lengths it will take to get 400 $13\frac{5}{8}$ in. pieces.
	●	To find the number of pieces that can be cut from each 8-ft length, some students will divide. Others may do repeated subtraction and some may draw a picture and perhaps use repeated addition until they get close to 8 ft. Once they know how many pieces come from each 8-ft length, they can again divide to find out how many of the 8-ft lengths are needed to get 400 pieces.
Solve It	○	58 8-ft pieces of stock
		Here are two possible solution paths:
		Students may use division. First find out how many $13\frac{11}{16}$ in. kit pieces can come out of 8-ft stock. Convert 8 ft to inches; $8 \times 12 = 96$ in. $96 \div 13\frac{11}{16}$ is 7 with a small remainder. Since each 8-ft piece gives 7 pieces, to find how many 8-ft pieces are needed, students divide the number of kits by 7. $400 \div 7$ is just over 57. So students need 58 8-ft pieces of stock.
		Other students may use repeated subtraction to find the number of $13\frac{11}{16}$ pieces in 8 ft. They find that they can subtract $13\frac{11}{16}$ from 96 seven times. Since the next part is whole number division, they may finish the problem as in the path above.
Look Back	●	Discuss whether students think that an estimated answer is sufficient for a problem such as this. How many extra 8-ft boards may have been purchased if just an estimate is used? If they estimate 7 pieces can be cut from one board and $400 \div 7$ is approximately 60 because $60 \times 7 = 420$, then two extra boards might have been purchased.
	○	A way to check the solution is to solve by a different method. Encourage students to share their solution methods in small groups.

2 | USING FRACTIONS

Solving Problems Using Fractions

Rosemary's club is making birdhouse kits from 8-ft wood stock. The base of each birdhouse requires a piece of stock that is $13\frac{5}{8}$ in. long. Each cut consumes an additional $\frac{1}{16}$ in. of stock. How many 8-ft pieces of stock are needed to provide the bases for 400 birdhouse kits?

Find Out	○ What question must you answer?
	• What is a reasonable estimate?
Make a Plan	○ What information do you have? What information do you need?
	• How will you use the information to find the answer?
Solve It	○ Find the answer. Use your plan. Keep a record of your work.
Look Back	• How does your answer compare with your estimate?
	○ How could you check if your answer is right?

2 | USING FRACTIONS
Try It Out

1. Make a Plan To help students plan their solution, ask:

- What does the problem ask you to find? (Is there enough fabric? If so, how much material is left after $8 \times 1\frac{1}{3}$ yd is used?)

- What operations will you use to solve the problem? (Multiply $8 \times 1\frac{1}{3}$, then subtract that from $11\frac{1}{4}$.)

Solution Path

Yes, there is enough. There will be $\frac{7}{12}$ yd left over.

Find the number of yards used to make the mats. $8 \times 1\frac{1}{3}$ is $8 \times \frac{4}{3}$ or $\frac{32}{3} = 10\frac{2}{3}$ yd.

Find how much fabric is left over. Subtract $11\frac{1}{4} - 10\frac{2}{3}$ to get $\frac{7}{12}$ yd.

---

2. $\frac{3}{4}$ gal

$\frac{9}{2} \div 6 = \frac{9}{12}$ or $\frac{3}{4}$

Alternately, students may draw $4\frac{1}{2}$ gal divided into half gallons so there are 9 halves. They will put 6 of those aside and consider the 3 remaining halves. If they divide those in half to get fourths, there will be 6 fourths. Adding the halves and the fourths gives $\frac{1}{2} + \frac{1}{4} = \frac{3}{4}$.

3. $1\frac{5}{6}$ mi

If group two hikes half the distance that group one hiked, then they hiked $\frac{1}{2}$ of $3\frac{2}{3}$ which is $\frac{1}{2}$ of 3 plus $\frac{1}{2}$ of $\frac{2}{3}$ or $1\frac{1}{2} + \frac{1}{3} = 1\frac{5}{6}$. Alternately, students can multiply $\frac{1}{2} \times 3\frac{2}{3} = \frac{1}{2} \times \frac{11}{3} = \frac{11}{6} = 1\frac{5}{6}$.

---

4. 6 kinds

$\frac{2}{5} \times 15 = 6$

5. $\frac{1}{6}$ lb

If they ate $\frac{2}{3}$ of the raisins, then $\frac{1}{3}$ remained so $\frac{1}{3}$ of $\frac{1}{2} = \frac{1}{3} \times \frac{1}{2} = \frac{1}{6}$.

---

6. $1\frac{1}{2}$ and $1\frac{3}{8}$ for one pile. $1\frac{3}{4}, \frac{5}{16}$, and $\frac{13}{16}$ for the other.

Students may choose to **Guess and Check**, **Make an Organized List**, or **Use Objects** to solve this problem. Encourage students to share their solution strategies.

$1\frac{1}{2} + 1\frac{3}{8} = 1\frac{3}{4} + \frac{5}{16} + \frac{13}{16} = 2\frac{7}{8}$

7. a. Answers will vary.
Possible answers include $\frac{13}{12} > 1, \frac{12}{4} > 2$.

b. Answers will vary.
Possible answers include $\frac{36}{42} < 1, \frac{4}{16} < \frac{1}{2}$.

c. Answers will vary. A possible answer is $36 \div \frac{3}{4} = 48$ and $48 > 20$.

---

8. 1; $\frac{4}{9}$ of the number is $4 \times \frac{1}{4}$

☑ **Informal Assessment** Ask: If $\frac{1}{8}$ of a number is 10, what is $\frac{1}{2}$ of the number? (40; $\frac{1}{2}$ of the number is $\frac{4}{8}$ of the number. If $\frac{1}{8}$ of the number is 10, $\frac{4}{8}$ of the number is 4 times that or 40.)

9. $\frac{1}{2} + \frac{1}{4} + \frac{1}{8} + \frac{1}{16}$

One strategy is to select a fraction with a denominator of 16 and a numerator that makes a fraction that can be simplified to a unit fraction like $\frac{8}{16}$. Then subtract that from the $\frac{15}{16}$. Then from the remaining $\frac{7}{16}$ subtract another, say $\frac{2}{16}$ and continue until you have no sixteenths remaining.

2 | USING FRACTIONS

Try It Out

1. Eight campers decide to make padded mats to sit on. They have $11\frac{1}{4}$ yards of fabric. Each mat requires $1\frac{1}{3}$ yards of fabric. Is there enough fabric? If so, how much will be left?

2. At science camp, $4\frac{1}{2}$ gal of milk will be poured equally into 6 containers. What size containers should they use?

3. Two camp groups took a hike along a marked trail. The first group hiked $3\frac{2}{3}$ mi. The second group hiked half that distance. How far did the second group walk?

4. On the hike, Marta identified 15 different kinds of wildflowers. If $\frac{2}{5}$ of them were white, how many kinds of white wildflowers did she find?

5. A half pound of raisins was in a jar in Wen's cabin. She and her cabin mates ate $\frac{2}{3}$ of them. How much was left?

6. The boys in Jamil's cabin found a log. They decide to make a bench by using books to support the log. The thicknesses of the books are $1\frac{1}{2}$, $\frac{13}{16}$, $\frac{5}{16}$, $1\frac{3}{8}$, and $1\frac{3}{4}$ inches respectively. How could they arrange the books to have two piles the same height?

7. Use 3, 4, 7, 9, 12, 13, 15, 16, 36, 42 to do the following.

a. Write a fraction that is > 1. That is > 2.

b. Write a fraction that is < 1. That is $< \frac{1}{2}$.

c. Write a division sentence containing a fraction that has a quotient > 20.

8. If $\frac{1}{9}$ of a number is $\frac{1}{4}$, what is $\frac{4}{9}$ of the number?

9. A unit fraction has a numerator of 1. Write $\frac{15}{16}$ as the sum of unit fractions each having a different denominator.

2 USING FRACTIONS

Stretch Your Thinking

1. The information about Harriet Tubman came from *The Story of Harriet Tubman, Conductor of the Underground Railroad* by Kate McMullan, Bantam Doubleday Dell Books for Young Readers ©1991. You may also find information about Harriet Tubman at www.camalott.com\~rssmith\Moses.html and other websites. Please note that data from different sources does not agree about birth date or exact numbers of years for events.

Find Out To help students understand the problem, ask:

- What information does the problem give you? What do you need to find out? (The problem splits Harriet's life up into years and fractions. $\frac{7}{23}$ life + $\frac{3}{23}$ life + $\frac{1}{23}$ life + 3 years + $\frac{5}{23}$ life + 25 years = whole life. What is missing is the length of Harriet's life. Once you know how long Harriet's life was, what else do you need to do? (Find $\frac{7}{23}$ of her life to find out how old she was when she escaped to freedom.)

- About how many years did Harriet live? How do you know? (Close to 100 because 25 years is almost $\frac{1}{4}$ of her life.)

Solution Paths

a. Harriet was 92 when she died.

Students know that $\frac{7}{23}$ life + $\frac{3}{23}$ life + $\frac{1}{23}$ life + 3 years + $\frac{5}{23}$ life + 25 years = whole life. So $\frac{16}{23}$ life + 28 is the number of years that Harriet lived. That means that 28 is $\frac{7}{23}$ the number of years that Harriet lived. So $\frac{1}{23}$ is 4 years. Harriet lived $\frac{23}{23}$ or 23 × 4 or 92 years.

b. Harriet was 28 when she escaped to freedom.

$\frac{1}{23}$ of Harriet's life is 4 years, and she was a slave for the first $\frac{7}{23}$ of her life. 7 × 4 = 28.

2. 426 or 427 students depending on whether they stretch out a little or get closer together

The perimeter of the middle school is 2 × 12 × 240 + 2 × 12 × 520 or 18,240 in. Divide 18,240 by $42\frac{3}{4}$ to find out how many students it takes to surround the perimeter.

✓ **Informal Assessment** Ask: If King Middle School is in a city that is 2 miles square, how many hand-holding citizens would it take to surround the city? Assume that the average citizen takes up $42\frac{3}{4}$ in. (11,856 or 11,857 citizens; The perimeter of the city is 8 mi. There are 5,280 ft or 63,360 in. per mile. The perimeter of the city is 506,880 in. Divide by $42\frac{3}{4}$ or 42.75 to find the number of citizens it takes to surround the city.)

3. a. 54

When he became the 43rd president, G.W. Bush was $\frac{27}{38}$ his father's age and his father was 22 when G.W. was born. That means that students need to look for a fraction that is equivalent to $\frac{27}{38}$ that has a denominator that is 22 more than the numerator. List fractions equivalent to $\frac{27}{38}$ to find $\frac{54}{76}$. The numerator is G.W. Bush's age.

b. 42

G.W. Bush was $\frac{21}{32}$ his father's age when his father became the 41st president. Look for a fraction equivalent to $\frac{21}{32}$ where the numerator is 22 less than the denominator. That fraction is $\frac{42}{64}$. The numerator is G.W. Bush's age.

2 USING FRACTIONS

Stretch Your Thinking

1. Harriet Tubman, a courageous leader of the Underground Railroad, led nineteen missions in which she helped over three hundred slaves escape to freedom. She spent the first $\frac{7}{23}$ of her years as a slave. Then she led slaves to freedom for the next $\frac{3}{23}$ of her years. Next she spent $\frac{1}{23}$ of her years working for the Union as a scout, spy, and nurse during the Civil War. After the war, she continued to nurse people who needed help for 3 years. She spoke about her experiences on the Underground Railroad while she was married to Nelson Davis for the next $\frac{5}{23}$ of her years. She lived the next, and last, 25 years, close to $\frac{1}{4}$ of her years, as a widow.

   a. How old was Harriet Tubman when she died?

   b. How old was she when she escaped to freedom?

Slaves sometimes made quilts that were really secret maps to freedom. The quilt contained pictures and landmarks to help guide the slaves to a safe place. These safe places were part of the Underground Railroad.

2. The students at King Middle School want to organize a tribute to Martin Luther King, Jr. They think it would be symbolic to make a line of people with joined hands. They figured that the average middle school student is $42\frac{3}{4}$ in. wide in the hand-holding position. King Middle School is on a rectangular lot that is 240 ft by 520 ft.

   How many participants do they need to line the perimeter of the school with hand holders?

3. When his father, George Bush, became the 41st president, George W. Bush was $\frac{21}{32}$ his father's age. When G.W. Bush became the 43rd president, he was $\frac{27}{38}$ his father's age.

   a. If George Bush was 22 when G.W. was born, how old was G.W. Bush when he became the 43rd president of the United States?

   b. How old was G.W. when his father became the 41st president?

2 | USING FRACTIONS

Challenge Your Mind

1. Make a Plan To help students decide how to proceed to solve the problem, ask:

- How will you use the information you have to solve part **a**? (Possible answer: The first bounce rises to $\frac{1}{3}$ of the 12 ft. The second bounce rises to $\frac{1}{3}$ of the first. The third rises to $\frac{1}{3}$ of the second. So the third is $12 \times \frac{1}{3} \times \frac{1}{3} \times \frac{1}{3}$.

- For part **b** where will you go to test the balls? (Students will need to find someplace where they can drop the ball.)

- How high should the drop point be? Could it be lower? (Since the final bounce may be $\frac{1}{27}$ or less of the initial bounce, the lowest the drop point could be would be 27 times the smallest bounce that can be accurately measured. If the ball was dropped from 27 in. the third rebound might be 1 in. Higher than 27 in. would be better because it would be easier to measure the third bounce. Students might be successful with a drop from about 6 ft.)

- What data do you need to record? (The kind of ball, the drop height, and the height of the third bounce.)

Solution Paths

a. $\frac{4}{9}$ ft or $5\frac{1}{3}$ in.

$12 \times \frac{1}{3} \times \frac{1}{3} \times \frac{1}{3}$ is $12 \times \frac{1}{27}$ or $\frac{12}{27}$ or $\frac{4}{9}$ ft
$\frac{4}{9}$ ft is $\frac{4}{9} \times 12$ or $\frac{16}{3}$ or $5\frac{1}{3}$ in.

b. Answers will vary.

c. Answers will vary.

---

2. One possible answer:

$$\frac{\boxed{4}}{\boxed{5}} + \frac{\boxed{3}}{\boxed{6}} - \frac{\boxed{1}}{\boxed{2}} = \frac{4}{5}$$

Students may notice that they can make $\frac{4}{5}$ from two of the 6 digits they have. Then all they have to do is find two other fractions that equal each other so that they can add and subtract the same amount. Another example is $\frac{4}{5} + \frac{1}{3} - \frac{2}{6}$.

☑ **Informal Assessment** Ask: Use the digits 2, 3, 4, 5 to make two fractions that when added together equal $3\frac{1}{4}$. ($\frac{3}{4} + \frac{5}{2} = \frac{13}{4} = 3\frac{1}{4}$)

2 | USING FRACTIONS

Challenge Your Mind

1. A ball dropped onto a hard surface rebounds exactly one-third the height from which it falls and continues to bounce one-third the height of each successive bounce.

a. How far will it rebound after hitting the surface for the third time if it is initially dropped from a height of 12 ft and continues to rebound one-third the height after each bounce?

b. Is there really a ball that rebounds $\frac{1}{3}$ the height from which it falls? Test several kinds of balls to find out. Tell the rebound fraction for each ball tested.

c. Choose a landmark such as the St. Louis Arch or the Eiffel Tower. Predict how high each ball you tested would rebound after the initial drop (based on your answers to **b**) if you dropped it from the top of this landmark. You will have to research to find the height of your chosen structure.

2. Use the digits 1, 2, 3, 4, 5, 6 to fill in the squares below to make a true equation. Use each digit exactly once.

$$\frac{\square}{\square} + \frac{\square}{\square} - \frac{\square}{\square} = \frac{4}{5}$$

2 | USING FRACTIONS

Wrap It Up

Fraction Flags

Discuss the problem with the class before they begin to work on their own. Ask:

- What does the problem ask you to do? (Make three very different flags on grid paper. Each flag has $\frac{1}{3}$ of one color, $\frac{1}{4}$ of a second, and the rest is the third color. What fraction of the flag is the third color? Explain how you know that each flag has the correct fractions of colors.)

- How can the grid paper help you? (You can figure out the total area of each color for each flag and compare it to the whole area of the flag to find the fractions.)

Solutions

The third color accounts for $1 - \frac{1}{3} - \frac{1}{4}$ of the flag or $\frac{12}{12} - \frac{4}{12} - \frac{3}{12}$ or $\frac{5}{12}$.

Flags and justifications of correct fraction make up will vary.

Possible Solution Paths

To find out the fraction for the third color of the flag, students may subtract the fractions of the other two colors from 1. Other paths include **Making a Picture or Diagram** of a very simple flag, and using that to find $\frac{5}{12}$.

Students should find the combined area of each color for each flag. They can use the grid paper to help. The students can then compare the area for each color to the whole flag to prove that the color is the right fraction of the whole flag.

Assessment Rubric

3 The student accurately identifies the fraction for the third color; draws three flags that are creative, complex, and significantly different, each composed of the correct fraction of each color; and clearly and accurately explains how to determine the fraction for each color using appropriate mathematics.

2 The student accurately identifies the fraction for the third color; draws three flags that are different, each composed of approximately the correct fraction of each color; adequately explains how to determine the fraction for each color; and uses appropriate mathematics however may have some arithmetic errors.

1 The student knows how to identify the fraction for the third color but calculates it incorrectly; draws three flags, each composed of three colors but fractional parts for some colors may be incorrect; and attempts to explain how to determine the fraction for each color but makes mathematical errors, or does not attempt to hit goal fractions.

0 The student may not identify the fraction for the third color or may not know how to calculate the fraction for the third color; makes fewer than 3 flags and may not have the correct fractions of colors; and does not successfully explain how to determine the fraction of each color.

2 USING FRACTIONS
Wrap It Up

Fraction Flags

Use grid paper to make three flags.

Each flag should look as different from the other two as possible. Your flags should be creative and contain some complex shapes—that is, shapes other than squares, rectangles or right triangles. Each flag should have three colors — $\frac{1}{3}$ of one color, $\frac{1}{4}$ of a second color, and the remainder the third color.

What fraction of each flag should be the third color?

For each flag, provide an explanation of how you know that the colors represent the required fraction of the flag.

This section focuses on solving problems using decimals. Students will develop increased computational fluency as they encounter a variety of multiple step problems.

Understanding Decimals

As students prepare to solve problems using decimals they will need to know how to round to the nearest tenth and nearest hundredth. They should understand how to add, subtract, multiply, and divide with decimals.

The problems in this section sometimes require students to convert from one measure to another within measurement systems. Students will also use rates such as miles per hour and find dimensions, area, surface area, and volume of regular figures.

You play an important role in helping your students choose an appropriate computational method. Students will sometimes benefit from using calculators to solve problems. When students use calculators to uncover patterns, calculator use helps them focus on the pattern rather than on the steps of the algorithm. Other times, doing the arithmetic by hand will reinforce students' mastery of the operations. The key to the difference is usually whether repetitive calculations are necessary to solve the problem or if mathematical reasoning yields just a few calculations.

It takes Mark 15.3 seconds to run up and down the bleachers. During this time his heart beats 126 times every minute. How many times does Mark's heart beat while he runs the bleachers 12 times, assuming that he is able to maintain a consistent running rate?
To solve this problem, students should:

- determine that this problem has multiple steps.

- realize that there are alternate paths to the solution –they may find the number of minutes that Mark runs and multiply that by the number of beats per minute, or alternatively they may find the number of seconds he runs and multiply that by the number of beats per second.

- know how to express seconds as a decimal part of a minute, or alternatively, know how to convert heart beats per minute to beats per second.

Five students rated movies on a scale of 1 to 5. To come up with a final rating, they throw out the high and low score and find the average of the remaining scores. The five scores for *Crazy Creeps* were 3.5, 4.75, 3.25, 4.5, and 4.2. What was the final rating for *Crazy Creeps*?
In a problem such as this, students must:

- be able to order decimals from smallest to greatest in order to throw out the high and low scores.

- know how to calculate a mean average using decimal numbers.

- be able to add and divide with decimals.

As students work with decimals, ask these key questions, as needed:

- What mathematical operations are needed to solve the problem?

- What rounded numbers can you use to mentally calculate a reasonable estimate? What is a reasonable estimate of the answer?

- Does the problem have multiple steps?

Solving Decimal Problems

Your students will be more successful if they can apply logical reasoning and combine general problem-solving strategies with methods specific to solving problems with decimals. Refer to *page vi* for a discussion of strategies.

Solving a Problem with *Guess and Check* One way to solve a problem is to estimate the answer, and check by doing the arithmetic to see how close it is to the target. Then use that result to choose a better estimate, and continue this process to get closer to the correct number.

Example: What is the length of the side of a cube with a volume of 10 cm³?

The volume of a cube is side³. The length of the edges of the cube can be found by finding the number which when cubed will result in 10 cm³. Students can estimate that it is a number between 2 and 3, much closer to 2 than 3. They will find that 2.1³ is 9.261 and 2.2³ is 10.648 so they will try a number between 2.1 and 2.2. As they can continue this process they will find that the side is between 2.15 and 2.16.

Solving a Problem by *Making It Simpler* and *Make a Picture or Diagram* Using compatible numbers to simplify a problem may help students develop a plan for solving the problem. Note: Compatible numbers are numbers that make mental math simpler.

Example: How many 1.5 cm tiles will it take to edge a square hot tub with a 4.8 m perimeter? Assume there is no space between the tiles.

Simplify the problem by thinking that each tile is 2 cm wide. Draw a picture to show how the tiles fit around a square with sides of 6 cm.

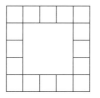

The project will require the number of tiles that fit along one edge, times four, plus four tiles for the corners.

The perimeter of the tub is 4.8 m so one side is 4.8 ÷ 4 or 1.2 m. 1 m is 100 cm, so 1.2 m is 120 cm. Each tile is 1.5 cm, so there are 120 ÷ 1.5 = 80 tiles per side. It takes 80 × 4 + 4 or 324 tiles to edge the hot tub.

Assessment

☑ **Informal Assessment** A suggestion for informal assessment will be found on each *Try It Out, Stretch Your Thinking,* and *Challenge Your Mind* page. The recommended question will help focus students' attention on one part of the problem-solving process.

Assessment Rubric An assessment rubric is provided for each *Wrap It Up.* Students' completed work may be added to their math portfolios.

Thinking About Decimals

Students use their number sense and estimation skills in these problems as they prepare for solving problems with decimals. Present one problem a day as a warm-up. You may choose to read the daily problem aloud, write it on the board, or create a transparency.

1. Write this problem and the digits of its answer on the board. Do not place a decimal point in the sum.

    13.458 + 2.442 = 1 5 9

   Say: **Using estimation, place the decimal point in the answer.**
   (15.9; One way to estimate the answer is to think that 13 + 2 is 15.)

2. Write this problem and the digits of its answer on the board. Do not place a decimal point in the answer.

    85.8 ÷ 5.5 = 1 5 6

   Say: **Using estimation, place the decimal point in the answer.**
   (15.6; One way to estimate the answer is to think that 85 ÷ 5 is about 100 ÷ 5 or 20.)

3. Write this problem and the digits of its answer on the board. Do not place a decimal point in the answer.

    0.84 ÷ 0.8 = 1 0 5

   Say: **Using estimation, place the decimal point in the answer.**
   (1.05; One way to estimate the answer is to think that 0.8 = 0.80, and there is a little more than one 0.80 in 0.84.)

4. Write this problem and the digits of its answer on the board. Do not place a decimal point in the answer.

    4.234 × 0.49 × 25.25 = 5 2 3 8 5 1 6 5

   Say: **Using estimation, place the decimal point in the answer.**
   (52.385165; One way to estimate the answer is to think that 0.49 is about one-half and half of 4 is 2 and 2 × 25 is 50, so the answer is around 50.)

5. **Sam divided 80 by a mystery number and got 3.25 for an answer. Is the mystery number greater than or less than 80? How do you know?**
   (Less than, because there is more than one of them in 80.)

3 USING DECIMALS

Solving Problems Using Decimals

In this lesson, students will use the information provided in the problem and their own knowledge to solve the problem. They will find it helpful in developing their problem-solving plan to use compatible numbers to estimate the solution.

Using the Four-Step Method

Find Out	○ The first problem asks students to find out how far Earth travels during one revolution around the Sun. The second problem requires students to figure how many times the Moon orbits Earth during one revolution around the Sun. ● Students may first think of a simpler situation such as 40 mph for 2 days could be solved by multiplying $40 \times 24 \times 2$. Then by rounding 67,000 mph to 100,000 and 365.2 days to 400 and 24 hours to 20, students may estimate that Earth travels fewer than $100,000 \times 20 \times 400$ or 800,000,000 miles around the Sun. They might also estimate that since 27.3 days is a bit less than a month, the Moon orbits Earth a little more than 12 times during Earth's revolution around the Sun. Accept all estimates that can be mathematically justified.
Make a Plan	○ Students know that Earth travels at 67,000 mph for 365.2 days. They also know that the Moon orbits Earth every 27.3 days. They need to supply the information that there are 24 hr in each of those 365.2 days. ● For the first question, students need to multiply the rate of speed by the number of hours in a day by the number of days that Earth travels in one revolution. Some students may choose a variation of that such as finding out the distance traveled in a day, then multiplying that by the number of days. For the second question, students will divide the number of days for Earth's revolution of the Sun by the number of days for Moon's orbit around the Earth. Some students may choose a variation such as adding 27.3 until they reach 365.2 or subtracting 27.3 from 365.2 repeatedly, however, division is the most, straightforward method.
Solve It	○ 587,241,600 miles around the Sun and 13.4 orbits around the Earth.
Look Back	● Have students compare their results with their estimates. With such large numbers it may be difficult for students to assess the reasonableness of their estimates. It may help students to determine numbers that the answer must be between. In this problem, if they had estimated $50,000 \times 400 \times 20$ to get 400,000,000 mi, they would know that 587,241,600 is reasonable because it is between 400,000,000 and 800,000,000. ○ A way to check the solution is to solve by a different method.

3 USING DECIMALS

Solving Problems Using Decimals

Earth travels around the Sun at 67,000 miles per hour. It takes 365.2 days for Earth to make a complete revolution around the Sun. How far does Earth travel in that revolution?

The Moon takes 27.3 days to orbit Earth. How many times does the Moon orbit Earth during one of Earth's revolutions around the sun?

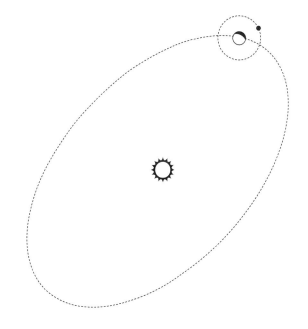

Find Out	○ What questions must you answer?
	• What are reasonable estimates?
Make a Plan	○ What information do you have? What information do you need?
	• How will you use the information to find the answer?
Solve It	○ Find the answer. Use your plan. Keep a record of your work.
Look Back	• How does your answer compare with your estimate?
	○ How could you check your solution?

3 | USING DECIMALS
Try It Out

1. **Make a Plan** To help students plan their solution, ask:

- What does the problem ask you to find? (How much less 124.3 in. is than 133.9 in., how much more 6.5 in. is than 5.27 in., how much more rain Palm Springs needs to reach 5.27 in. if it already has 4.72 in. + 0.3. in., and how many times greater 133.9 is than 5.27.)

- What words help you decide the operation to use in solving problems of comparison? (There are different kinds of comparisons. The words *how much more* or *how much less* are clues to the kind of comparison that you are being asked to make. This type of comparison suggests finding the difference or using subtraction. In the first two parts, you have to subtract to find how much more or how much less. In the third part, you have to add the additional rainfall to the initial rainfall to get the total rainfall so far, then subtract that from the average rainfall to find how much more. The words *how many times* in part **d** give a clue that this is a different kind of comparison. These words suggest using either multiplication or division, and in this particular case, division.)

Solution Path

a. 9.6 in. less than average

$$133.9 - 124.3 = 9.6$$

b. 1.23 in. more than average

$$6.5 - 5.27 = 1.23$$

c. 0.25 in. more

$$5.27 - (4.72 + 0.3) = 0.25$$

d. 25.4 times more

$$133.9 \div 5.27 = 25.4$$

2. a. $9.64 \times 8.75 = 84.35$

Most students will use the **Guess and Check** strategy for this problem. Using a calculator allows them to focus on the effect of moving digits to different positions rather than on the arithmetic.

b. $7.598 \times 4.6 = 34.9508$

☑ **Informal Assessment** Ask: Can you use the digits 1, 2, 3, 4, 5, 6 and one operation to make an expression that is approximately equal to 2? (Possible answers: $6.45 \div 3.21 = 2.01$, $4.35 \div 2.16 = 2.01$.)

3. a. 0.54 s

$$11.29 - 10.75 = 0.54$$

b. 0.37 s

$$11.12 - 10.75 = 0.37$$

Some students may need to be reminded that the faster the runner is, the shorter the running time will be.

3 | USING DECIMALS

Try It Out

1. The average annual rainfall for Limon, a tropical city on the east coast of Costa Rica, is 133.9 in. One recent year, Limon had only 124.3 in. of rain. The average annual rainfall for Palm Springs, California, is 5.27 in. Recently they had 6.5 in. of rain for the year.

   a. How much less annual rainfall than average did Limon recently have?

   b. How much more rainfall than average did Palm Springs recently have?

   c. Suppose that Palm Springs has already recorded 4.72 in. of rain this year. A weather system dumps 0.3 in. more. How much more rain is needed to reach the average amount?

   d. The average annual rainfall of Limon is how many times the amount for Palm Springs?

2. The equation 9.8 × 7.654 = 75.0092 has each of the digits 4, 5, 6, 7, 8, and 9 exactly once on the left-hand side. The right-hand number is close to but less than 80.

   a. Can you find another multiplication equation that contains each digit 4, 5, 6, 7, 8, and 9 exactly once on the left-hand side and has the greatest possible product less than 85?

   b. Can you use the same digits on the left to find the greatest product less than 35?

3. Eight women ran in the 2000 Olympic finals of the women's 110-meter dash.

Name/Country	Seconds
Marion Jones (USA)	10.75
Ekaterini Thanou (Greece)	11.12
Tanya Lawrence (Jamaica)	11.18
Merlene Ottey (Jamaica)	11.19
Zhanna Pintusevych (Ukraine)	11.20
Chandra Sturrup (Bahamas)	11.21
Sevatheda Fynes (Bahamas)	11.22
Debbie Ferguson (Bahamas)	11.29

   a. What is the difference in time between first and last place?

   b. What is the difference in time between first and second place?

3 | USING DECIMALS
Stretch Your Thinking

1. Find a Plan To help students understand the problem, ask:

- What does *averaged* mean? (Although there are different kinds of averages, this refers to a mean average. So you add the scores, then divide by the number of scores.)

- How many scores will you average for each piece of equipment? (3)

- What questions does this problem ask you to answer? (Who has the highest all-around score when adding the average for each event and what are their all-around scores?)

- What information do you have? What information do you need? (You know the five scores for each apparatus. You need the average of the middle three scores for each event. Then you need to add the averages for the four events to get a total score.)

Solution Paths

Kristal has the higher score by almost seven-tenths of a point. Morgan's score is 36.10. Kristal's score is 36.77.

Making an Organized List of the scores is helpful for this problem because it allows students to easily eliminate the highest and lowest scores, and allows them to group similar numbers for adding.

Morgan's all-around score is:

$[(8.9 + 8.9 + 9.0) \div 3 + (9.5 + 9.5 + 9.5) \div 3 + (9.0 + 9.0 + 9.1) \div 3 + (8.6 + 8.6 + 8.7) \div 3]$ or 36.10

Kristal's all-around score is:

$[(8.6 + 8.6 + 8.8) \div 3 + (9.4 + 9.6 + 9.6) \div 3 + (8.8 + 8.9 + 8.9) \div 3 + (9.6 + 9.7 + 9.8) \div 3]$ or 36.77

---

2. 76 times taller

Students will need to convert centimeters to meters or meters to centimeters. You may wish to make copies of the conversion chart found on the bottom half of page 74.

Some students may use the **Make it Simpler** strategy by thinking that if the adult was 25 cm tall and the kit 2.5 cm, the adult is 10 times as tall as the kit. That is, $25 \div 2.5$ is 10. Then they will follow the same procedure using the numbers in the problem.

1.9 m is 190 cm since there are 100 cm in a meter. $190 \div 2.5$ is 76.

3. 6.9 miles, $1,199.30

$1,000 \div 145.37$ is 6.9 mi
145.37×8.25 is $1,199.30

☑ **Informal Assessment** Ask: If Lian earns $150.25 per mile, will he have to run farther than he does in the original problem to make $1,000? How far will he have to run? How much more or less is that than the number of miles to run to make $1,000 in the original problem? (No, 6.7 mi, 0.2 mi; He has to run $1,000 \div $150.25 = 6.7$ mi and that is $6.9 - 6.7 = 0.2$ mi less.)

---

4. $19.25 more per ounce

Find the cost per ounce for both moisturizers. The drug-store brand costs $8.95 for 12 oz, so 1 oz is $8.95 \div 12$ or about $0.746 per ounce. The name brand cost $34 for 1.7 oz, so 1 oz is $34 \div 1.7$ or $20 per ounce. The question asks how much more, so compare the two prices, $20 - $0.75 is $19.25 more per ounce.

5. a. $5.66

$22.64 \div 4 = $5.66

b. $6.63

$22.64 - $2.75 = $19.89
$19.89 \div 3 = $6.63

3 | USING DECIMALS

Stretch Your Thinking

1. In competitive gymnastics, four to six judges give scores on each apparatus. The high and low scores are discarded and the remaining scores are averaged. A gymnast's all-around score is compiled by adding each of the scores for the individual events. Here are scores for each piece of apparatus for two competitors. Who has the highest all-around score? What are their all-around scores?

Morgan
Vault: 9.2, 8.9, 8.8, 8.9, 9.0
Uneven bars: 9.5, 9.5, 9.4, 9.7, 9.5
Balance beam: 8.9, 9.0, 9.1, 9.1, 9.0
Floor exercise: 8.6, 8.9, 8.7, 8.6, 8.5

Kristal
Vault: 8.5, 8.6, 8.6, 8.8, 8.9
Uneven bars: 9.6, 9.3, 9.4, 9.6, 9.7
Balance beam: 8.9, 8.8, 8.7, 8.9, 9.2
Floor exercise: 9.8, 9.7, 9.8, 9.5, 9.6

2. A kangaroo kit is 2.5 cm long at birth. A fully-grown kangaroo is 1.9 m tall. How many times taller is an adult than a kit?

3. Lian is in a jogathon to raise money for charity. Friends have pledged a total of $145.37 per mile. How far does he have to run to earn $1,000? He was actually able to run 8.25 mi. How much did he earn for charity?

4. A dermatologist compared the ingredients and effectiveness of a drug-store moisturizer and a brand-name cosmetic-counter moisturizer. The ingredients and effectiveness are similar for both brands. The drug-store product costs $8.95 for 12 oz and the brand-name product costs $34.00 for 1.7 oz. How much more per oz does the more expensive product cost?

5. Pedro and three friends go out to lunch. The bill is $22.64.

a. If they split the bill evenly, how much will each person pay?

b. Everyone, except Pedro, got items that cost about the same. Pedro got only a $2.75 cup of soup. If Pedro pays only $2.75, how much will each of his friends have to pay?

3 | USING DECIMALS
Challenge Your Mind

1. **Find Out** To help students understand the problem, ask:

- If the length of the side of the larger square is 4, what is the diameter of the circle? (4)

- How do you find the area of a circle? ($A = \pi r^2$ or $\pi \times$ the square of the radius.)

- What can you tell about the smaller square? (The diagonal of the small square is 4. If both diagonals are drawn, they form four isosceles right triangles that have legs that are each 2 units long.)

Solution Paths

a. 3.44 or 3.4 sq units

The large square has an area of 4×4 or 16. The circle has an area of 3.14×2^2 or 12.56 sq units. So the large square has $16 - 12.56$ or 3.44 sq units more area than the circle.

b. 4.56 sq units

The area of the small square is the sum of the four right triangles formed by the diagonals. The area of each of these isosceles right triangles is 2 square units ($A = \frac{1}{2}bh$ or $\frac{1}{2} \times 2 \times 2 = 2$). Therefore, the area of the square is 4×2 or 8 square units. The circle has an area of 12.56 square units so the area of the circle outside of the small square is $12.56 - 8$ or 4.56 sq units.

c. Two times.

The large square is 16 sq units, the small square is 8 sq units. 16 is 2×8.

---

2. a. Answers will vary. One possible set of dimensions is 8 cm $\times$ 4.5 cm $\times$ 6.3 cm. This box will have a surface area of 229.5 cm².

If one of the three dimensions of the box is 8 cm, and the volume is 225.8 cm³, the two remaining dimensions have a product equal to $225.8 \div 8$ or 28.225. Students may **Make a Table** or **Guess and Check** to find two numbers that produce that product. One possible pair is 4.5 and 6.3 (these are rounded so they do not give the exact product). The surface area of that box would be $2 \times 8 \times 4.5 + 2 \times 8 \times 6.3 + 2 \times 4.5 \times 6.3$ or 229.5 cm². Other pairs of numbers that produce the product 28.225 will produce different surface areas.

b. A cube 6.09 on a side; a cube 6.09 on a side has a volume of about 225.8 cm³. Students may round to sides of 6.1 cm. The surface area is about 222.5 cm² (or 223.3 cm² if 6.1 cm is used).

Students may choose a simpler volume to find that the cube uses the least amount of material thus having the smallest surface area. Example: If the volume is 8, $1 \times 1 \times 8$ has a surface area of 34 cm²; $1 \times 2 \times 4$ has a surface area of 28 cm²; and $2 \times 2 \times 2$ has a surface area of 24 cm². If the box is a cube, the length of each side is the cube root of 225.8 cm³. Students may use a calculator to find the cube root of 225.8 or may use **Guess and Check** to find it. Since $6^3 = 216$, the cube root of 225.8 is slightly larger than that. Then since $6.1^3 = 226.981$, the cube root must be between 6 and 6.1, closer to 6.1 than to 6. Try 6.09. That gives a volume of 225.87 cm³. The surface area is $6 \times 6.09 \times 6.09$ or 222.5 cm².

✓ **Informal Assessment** Ask: Name the dimensions of two different rectangles with an area of 16.2 cm². (Possible answers: 8.1 cm by 2 cm and 4.5 cm by 3.6 cm. There are an infinite number of answers.)

3 | USING DECIMALS

Challenge Your Mind

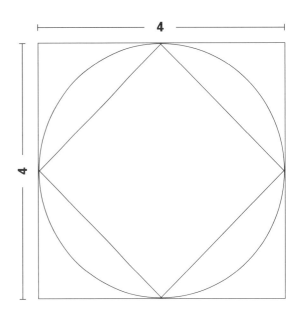

1. The figure shows a square within a circle, and the circle inside of a larger square. The larger square has a side length of 4 units.

a. What is the area of the portion of the larger square that is outside the circle?

b. What is the area of the portion of the circle that is outside the small square? (use $\pi = 3.14$)

c. The area of the large square is how many times the area of the smaller one?

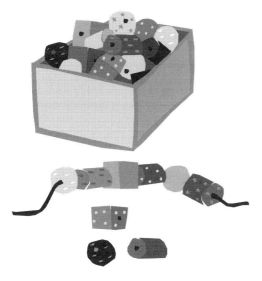

2. Suppose you wanted a box that would hold about 225.8 cm³ of beads.

a. If one of the dimensions of the box is 8 cm, what are possible measures for the other two dimensions? Using those dimensions, what is the surface area?

b. What are the dimensions of a box to hold the beads that would use the least amount of cardboard (i.e. give the smallest surface area)? What is the surface area of this box?

3 USING DECIMALS
Wrap It Up

Heels Over Head

Discuss the problem with the class before they begin to work on their own. To explain why the total score is multiplied by 0.6, use this example. Example: If there are five judges and the marks are 8, 7, 7, 7, 6.5 and the degree of difficulty is 2, then the score would be $21 \times 2 = 42$. If there are seven judges and the marks are 8, 7, 7, 7, 7, 7, 6.5 and the degree of difficulty is 2, then the score would be $35 \times 2 = 70$. The scores for both dives were the same except for the number of judges. To even up the five scores (7 judges) with the three scores, multiply the seven-judge score by 0.6. In this example, the scores for both dives are now the same, 42 (because $70 \times 0.6 = 42$). Ask:

- What does it mean to drop the highest and lowest marks? (Put the marks in order from least to greatest and remove the first and last marks, leaving the five middle marks.)

- What do you have to find for **Part A**? (Rodriguez's score.)

- What do you have to find for **Part B**? (A set of seven marks between 7.0 to 8.5 that yields a score that is 2 or fewer points less than Rodriguez's score. A second set of seven marks between 7.0 to 8.5 that yields a score 2 or fewer points greater than Rodriguez's score.)

Solutions
Part A 72.24

Part B Answers will vary. One solution for the lower score is 7.0, 7.0, 7.5, 7.5, 7.5, 7.5 and 8.5. One solution for the higher score is 7.0, 7.5, 7.5, 7.5, 8.0, 8.0, and 8.5.

Solution Paths
Part A Discard 9.0 and 8.0.
$(8.5 + 9.0 + 8.5 + 8.5 + 8.5) \times 2.8 \times 0.6$ is 72.24.

Part B Students may **Work Backward** to solve this problem. 2 points fewer than 72.24 is 70.24. Divide by 0.6 and the 3.2 difficulty to get a raw score of 37.6. Divide by 5 to get the average mark of 7.52. Try five 7.5 marks to see if the total score is 2 or fewer points than 72.24. $7.5 \times 5 \times 3.2 \times 0.6$ is 72 points. $72.24 > 72 > 70.24$. To have seven marks, students will need to add a high and a low mark between 7.0 to 8.5.

To find the winning set of scores, students may **Work Backward** in the same way or may use **Guess and Check**. One possible answer is 9.0, 7.7, 8.5, 7.7, 7.7, 7.6, 7.8. This gives a total score of 73.92 which is 1.68 points more than Rodriguez.

Assessment Rubric
3 The student accurately chooses the high and low score to discard; adds and multiplies correctly to find the final scores; and finds two sets of marks that yield scores within the range specified by the problem.

2 The student accurately chooses the high and low score to discard; adds and multiplies with minor errors; and finds two sets of marks that yield scores within the range specified by the problem.

1 The student chooses the wrong marks to discard; makes arithmetic errors; and does not find both sets of marks to yield scores within the range specified by the problem.

0 The student forgets to discard the high and low marks, makes arithmetic errors, and does not attempt, or is not successful in repeated attempts, to find two sets of marks which give scores within the range specified by the problem.

3 | USING DECIMALS

Wrap It Up

Heels Over Head

There are seven judges in Olympic springboard diving competition. The judges rate each dive with a mark ranging from 0 to 10 in full-or half-point increments. The highest and lowest marks are dropped. The remaining marks are added together and multiplied by the degree of difficulty (1.2 to 3.6). This total is then multiplied by 0.6 in order to make the scores comparable to competitions with five judges. Combined semifinal and final scores determine final results.

Part A

Here are the marks Rodriguez received for his final dive. This dive has a 2.8 degree of difficulty.

8.5, 9.0, 8.0, 8.5, 9.0, 8.5, 8.5

What is his score?

Part B

Up to his final dive, Rodriguez was tied for first place with Jingjing. Jingjing's final dive has a 3.2 degree of difficulty. His marks on the final dive range from 8.5 to 7.0. Find a set of marks where Jingjing loses by less than 2 points. Find a set of marks where Jingjing wins by less than 2 points.

This section focuses on using percents to solve problems.

Understanding Using Percents

As students prepare to use percents to solve problems, they need to understand several basic percent concepts. It is helpful for them to be at ease with using several benchmark percents and their multiples such as 1%, 10%, 25%, 50%, $33\frac{1}{3}$%. They should know percent names for $\frac{1}{100}, \frac{1}{10}, \frac{1}{5}, \frac{1}{4}, \frac{1}{3}$, and $\frac{1}{2}$. They should know that the basic percent relationship is

$$\frac{Part}{Whole} = \frac{Rate}{100}.$$

Students should be familiar with finding percent of change. A recurring challenge in this section is choosing the appropriate part and whole to compare.

Steve has 8 coins that are worth a total of $1. 25% of the coins are quarters. 40% of the $1 is dimes. What coins does Steve have?
To solve this problem, students will need to:

- be able to find a percent of the given quantity and know whether that percent represents a number of coins or the value of the coins.

- use the information about the number of dimes and quarters and their value to figure out the value and number of the remaining coins.

Steve's stocks decreased 10%, then increased 20%. Rosa's stocks increased 20%, then decreased 10%. If their stocks were worth $1,000 to start, how much do they end up with? Why?
In a problem such as this, students should:

- know how to calculate a percent of increase and decrease.

- know the whole upon which the increase or decrease is based.

- examine the reason that the answer is the same for both people.

As students work with percents, ask these key questions, as needed:

- What is the whole? The part?

- What are you being asked to compare?

- Does the problem have multiple steps?

- Is there more than one answer?

Solving Percent Problems

Your students will be more successful if they can apply logical reasoning and combine general problem-solving strategies with methods specific to percent problems. Refer to *page vi* for a discussion of strategies.

Solving a Problem by *Making a Table* and *Guess and Check* When a problem contains a lot of data, it is often helpful to organize the data into a table. When students use the **Guess and Check** strategy, a table will help them keep track of the results of their guesses, and thus facilitate the choosing of increasingly better guesses.

Example: At Science Camp, students mixed two batches of trail mix. One mix was 5% chocolate chips. Another mix was 30% chocolate chips. The 5% mix didn't have enough chips, but the 30% mix melted and got too messy. They decided that a 15% mix would be ideal. They have 20 c of 30% mix and 50 c of 5% mix. How much 15% mix can they make, and how much of the 30% and 5% mixes will they use?

Make a chart with the information given in the problem, the information that can be determined from the initial information, and the data gathered from successive guesses.

Amount 5% mix	50 c	40 c	30 c
Amount chips from 5% mix	2.5 c	2 c	1.5 c
Amount 30% mix	20 c	20 c	20 c
Amount chips from 30% mix	6 c	6 c	6 c
Total amount new mix	70 c	60 c	50 c
Percent chips in new mix	8.5 ÷ 70 or 12.1%	8 ÷ 60 or 13.3%	7.5 ÷ 50 or 15%

The students use 30 c of the 5% mix, and 20 c of the 30% mix to get 50 c of the 15% mix.

Assessment

☑ **Informal Assessment** A suggestion for informal assessment will be found on each *Try It Out, Stretch Your Thinking,* and *Challenge Your Mind* page. The recommended question will help focus students' attention on one part of the problem-solving process.

Assessment Rubric An assessment rubric is provided for each *Wrap It Up*. Students' completed work may be added to their math portfolios.

Thinking About Percents

These problems will help students use their number sense and estimation skills as preparation for using percents to solve problems. Present one problem a day as a warm-up. You may choose to read the daily problem aloud, write it on the board, or create a transparency.

1. **State equivalent percents for these fractions and ratios:** $\frac{3}{4}, \frac{1}{2}$, **2 out of 5, and 2 out of 3.**
   (75%, 50%, 40%, $66\frac{2}{3}$%; these are benchmark percents. Students should know percent names for $\frac{1}{100}, \frac{1}{10}, \frac{1}{5}, \frac{1}{4}, \frac{1}{3}, \frac{1}{2}$ and their multiples.)

2. **What is a 15% tip on $10? A 20% tip? What is a 15% tip on $20? On $30? On $14?**
   ($1.50, $2.00, $3.00, $4.50, between $2 and $3.)
   Encourage students to mentally calculate tips using compatible numbers and rounding the amounts to the nearest half-dollar or dollar. A fifteen-percent tip is 10% plus half of that amount, so on $10 it is $1.00 plus $0.50 or $1.50. A twenty-percent tip is twice 10%, so on $10 it is 2 × $1.00 or $2.00.

3. **Suppose there were 10 club members last year and 15 this year. What is the percent of increase from last year to this year?**
   (50%; the increase from 10 to 15 is 5 and 5 ÷ 10 is 50%.)

4. **Suppose your brother lends you money at 10% interest per day, compounded daily. If he lends you $10 the day before your allowance, how much would you owe him on allowance day? On the day after allowance day?**
   ($11, $12.10; 10% of $10 is $1, so after 1 day, you owe $11. The second day you owe 10% based on the $11. 10% of $11 is $1.10, so on the second day you owe a total of $12.10.)

5. **Which is the greater amount, 25% of 50 or 50% of 25? Why?**
   (They are equal; 25% of 50 is 12.5. 50% of 25 is 12.5. It is the same because multiplication is commutative. $a \times b = b \times a$. 25% is 0.25 = 0.01 × 25 so 25% of 50 is 0.01 × [25 × 50] and 50% of 25 is 0.01 × [50 × 25].)

4 | USING PERCENTS

Solving Problems with Percents

In this lesson students will need to be able to identify the amount of increase or decrease and the starting amount in order to calculate the percent of increase or decrease.

Using the Four-Step Method

Find Out	
	○ The problem asks students to find out the percent of change for club funding and dance funding.
	● *Percent of change* is the ratio of *a* to *b* where *a* is the difference between the starting amount and the ending amount, and *b* is the starting amount. It is expressed as a percent. *Percent of change* can be an increase or a decrease.
	○ The change in both cases is $200. For clubs, $200 \div 550$ is about $\frac{2}{5}$ or 40%. Since the funding went down for clubs, that is a decrease of about 40%. For dances, $200 \div 400$ is $\frac{1}{2}$ or 50%. Dances got more money, so that is an increase of 50%.

Make a Plan	
	● Students know the amount taken from clubs and given to dances. They also know how much is budgeted for clubs after the budget cut, and how much money dances will now get. They need to figure out the initial budgets for clubs and dances in order to calculate the percent of change.
	○ First figure out the initial budgets for each by subtracting $200 from the dance budget and adding $200 to the club budget. Then find the percent that $200 is of the initial budgets for clubs and dances. Finally, identify the percent of change as an increase or decrease for each.

Solve It	
	● Clubs got a 36% decrease. Dances got a 50% increase. The initial budget for clubs was $350 + $200 or $550. The initial budget for dances was $600 − $200 or $400. $200 \div 550 = 0.364$. $200 \div 400 = 0.50$. Clubs got less money, therefore it is a decrease. Dances got more money, therefore it is an increase.

Look Back	
	○ Have students compare their results with their estimates. Ask students if their solution is reasonable when compared to their estimate?
	● A way to check the solution is to solve by a different method. Encourage students to share their solution methods in small groups.

4 USING PERCENTS

Solving Problems with Percents

When the student council revised its budget, they decided to increase the money for dances by $200. Since the amount of money in the budget remained the same, they decided to take the $200 from school clubs. School clubs will now receive $350 and dances will get $600. What percent of change was this for club funding? For dance funding?

Find Out	○ What questions must you answer?
	● What does *percent of change* mean?
	○ What are reasonable estimates?
Make a Plan	● What information do you have? What information do you need?
	○ How will you use the information to find the answer?
Solve It	● Find the answer. Use your plan. Keep a record of your work.
Look Back	○ How does your answer compare with your estimate?
	● How could you check if your answer is right?

4 USING PERCENTS

Try It Out

1. **Find Out** To help students understand the problem, ask:

- What does the problem ask you to find? (What percentage of all students voted for each type of food? Which, if any, of the foods got 50% of the vote?)

- How many students are in the school? How many students voted? (1,140; 296 + 474 + 75 + 73 or 918.)

Solution Path

a. 26% vote for pizza, 42% for tacos, 7% for quesadillas, 6% for hamburgers, and 19% did not vote.

There are 1,140 students at Chavez. 296 ÷ 1,140 or 0.2596 voted for pizza, 474 ÷ 1,140 or 0.4157 for tacos, 75 ÷ 1,140 or 0.0657 for quesadillas, 73 ÷ 1,140 or 0.0640 for hamburgers, and 222 ÷ 1,140 or 0.1947 did not vote. Students may not list the percent that did not vote, as the problem does not ask for this information.

b. Yes, tacos got over 50% of the vote.

918 students voted. 50% of 918 is 459, so the taco vote of 474 is more than 50%.

---

2. a. 11%; $\frac{124}{1,111} = 0.1116$

b. 14%; $\frac{156}{1,111} = 0.1404$

c. 75%; $1,111 - 280 = 831$, $\frac{831}{1,111} = 0.7479$

d. 33%; $370 ÷ 1,111 = 0.3330$

e. 14%; $51 ÷ 370 = 0.1378$

Some students may choose to list the numbers from 1 to 1,111 and count the numbers with each characteristic. A more efficient way is to **Make an Organized List** to complete these problems. The example below finds the numbers from 1 to 1,111 that have all even digits.

Numbers Considered	# of All Even
1 to 9 has 2, 4, 6, 8 for	4
10s, 30s, 50s, 70s, 90s have	0
20s has 20, 22, 24, 26, 28 for	5
Similarly 40s, 60s, 80s each have 5 for	15
100s, 300s, 500s, 900s, 1000s have	0
200s has 200 plus the same as 0 to 99	25
Similarly 400s, 600s, 800s each have 25 for	75
Total	124

3. Please note that asthma plans may vary. The percents and actions may vary depending upon the person.

a. Under 270; 60% × 450 is 270

b. No; 300 ÷ 480 = 0.625 or 62.5%

☑ **Informal Assessment** Ask: On one asthma plan, when the peak flow is between 60% and 80%, the person with asthma is supposed to take a puff of inhaled medicine. If the peak flow is 375 and the current flow is 325 should the person with asthma take a puff of medicine? (No, the current flow is 87% of peak flow, which is not within the range to take medicine.)

4 | USING PERCENTS

Try It Out

1. The 1,140 students at Chavez Middle School voted for what lunch they wanted for the last day of school. 296 voted for pizza, 474 voted for tacos, 75 voted for quesadillas, 73 voted for hamburgers, and the rest did not vote.

a. What percentage of all the students voted for each type of food?

b. Before the voting took place it was agreed that in order for a food win, it would need to get at least 50% of the votes, and it should be 50% of students who voted, not the entire school. Did any of the foods get enough votes to win? If so, which food?

2. What percentage of whole numbers from 1 to 1,111

a. have only even digits?

b. have only odd digits?

c. have both even and odd digits?

d. are multiples of 3?

e. are multiples of 3 with only odd digits?

3. Dharma has asthma. She uses a device called a peak-flow meter. It measures how fast she can move air out of her lungs. First, she uses the meter when she has no trouble breathing. This gives her *personal-best peak flow*. Then, when she is having trouble breathing, she measures the flow and compares that to her personal-best peak flow. If the flow measures below 60%, she is in her *danger zone* and is supposed to call her doctor.

a. What is the danger zone if the personal-best peak flow is 450 liters per minute?

b. Dharma's personal-best peak flow is 480 liters per minute. Today it is 300. Is she in her danger zone?

4 USING PERCENTS

Stretch Your Thinking

1. Make a Plan To help students plan their solution, ask:

- What question do you have to answer? (By what percent has the house increased? By what percent has the yard decreased?)

- What information is in the problem? What other information do you need to solve the problem? (The dimensions of the lot, the original house, and the new house are given. Students need to know how to find the areas of the lot, the original house, and the new house. The difference between the areas of the new house and the old house and the old yard and the new yard are also necessary.)

Solution Paths

The house increased by 140% and the yard decreased by 18%.

Students might **Make a Table** to organize the information.

	Dimensions	Area (ft²)	Compared to Original
Lot	150 ft × 75 ft	11,250	
Old House	25 ft × 50 ft	1,250	
Old Yard		10,000	
New House	40 ft × 75 ft	3,000	1,750 more
New Yard		8,250	1,750 less

Then they may find the percent increase for the area of the house $1,750 \div 1,250 = 1.4$ or a 140% increase. The percent decrease of the yard is $1,750 \div 10,000 = 0.175 = 17.5\%$ or an 18% decrease.

---

2. 80; Students may use **Guess and Check.** You may want to discuss strategies for mentally finding an increase of 25%. Encourage them to select numbers divisible by 4 so they can easily add $\frac{1}{4}$ of the number to the selected number. $80 \times 1.25 = 100$, 50% of 100 is 50. 50 is 30 less than 80.

☑ **Informal Assessment** Ask: Suppose Joe and Rosa were each making $10 per hour. Joe gets a 10% pay cut, and Rosa gets a 10% raise. After 6 months Joe gets a 10% raise and Rosa gets a 10% cut. Compare their last salaries to their beginning salaries. (They both make $9.90 per hour, 10¢ less than they started. That is a 1% decrease.)

3. 10 more; Students may use **Guess and Check** and **Make a Table** to find the number. First they will determine that 46.8% of 156 is 73 so Coretta made 73 free throws. Start by guessing that she makes 20 more shots. Keep guessing until the percent made is as close to 50% as possible. Each guess should lead to a better next guess.

More Shots	Total Shots	Made Shots	% Made
20	176	93	52.8%
10	166	83	50%

---

4. 120; 80% of 200 or 160 students like hamburgers. 25% of those are girls or $0.25 \times 160 = 40$ girls like hamburgers. Therefore, $160 - 40$ or 120 boys like hamburgers. Alternatively, students may reason that if 25% of the fans of hamburgers are girls, then 75% must be boys and 75% of 160 is 120.

5. No, she is charging $12.38, materials cost $12.50, so she is losing $0.12 per wreath.

55% of $25 is $13.75 so the sale price was $25 − $13.75 = $11.25. 10% of $11.25 is $1.13, so the new price is $11.25 + $1.13 or $12.38. 50% of $25 is $12.50. The cost of the materials ($12.50) is greater than the sale price ($12.38).

---

4 | USING PERCENTS
Stretch Your Thinking

1. In areas of the U.S. where the population is growing quickly and there is a limited amount of land, property values are soaring. Sometimes in desirable locations, people buy small houses on small lots and replace the small house with a much larger one. Suppose a lot is 150 ft × 75 ft and the original house was 25 ft × 50 ft. The new house is 40 ft × 75 ft. By what percent has the area of the house increased? By what percent has the yard area decreased?

2. A number is increased by 25%, then the resulting number is decreased by 50%. What was the original number if the final number is 30 less than the original?

3. Midway through the basketball season, Coretta figures that she has made 46.8% of her 156 free throws. How many free throws in a row does she have to make to bring her average to exactly 50%?

4. Eighty percent of the students in a sixth-grade class of 200, like hamburgers. 25% of the students who like hamburgers are girls. How many boys like hamburgers?

5. Marjorie makes Halloween decorations and sells them at a farmer's market. One item is a door wreath that she sells for $25. The first weekend of November, she decides to sell her wreaths at a discount. She marks them 55% off, but then realizes that she will be selling the wreaths for less than the cost of the materials to make them. She increases the discounted price by 10%. If the materials cost 50% of the original $25 price, will she cover her costs with the new price? Explain your answer.

4 | USING PERCENTS

Challenge Your Mind

1. Find Out To help students understand the problem, ask:

- What questions does the problem ask? (If you mix 20 qt of 25% lemon juice and 60 qt of 10% lemon juice, what percent of the resulting 80 qt mixture is lemon juice? How much of the 25% mixture and the 10% mixture could you put together to get the most 15% mixture possible?)

- How much lemon juice is in the 20 qt of 25% mixture? (5 qt lemon juice; $\frac{1}{4}$ of 20 is 5.)

Solution Paths

a. 13.8%

20 qt of 25% lemon juice contains $\frac{1}{4}$ of 20 or 5 qt lemon juice. 60 qt of 10% lemon juice contains $\frac{1}{10}$ of 60 or 6 qt lemon juice. The resulting 80 qt contains 5 + 6 or 11 qt lemon juice. $\frac{11}{80}$ is 0.138.

b. 20 qt of 25% lemon juice and 40 qt of 10% lemon juice. Students may **Make a Table** to record results of **Guess and Check**. Students who decide to use

all of the 10% mix will realize that even if they use all of the 25% mixture, the percent of lemon juice does not reach a 15% concentration. Therefore, they should consider using all of the 25% mixture and varying amounts of the 10% mixture.

Amount 25% mix	20 qt	20 qt	20 qt
Amount lemon juice from 25% mix	5 qt	5 qt	5 qt
Amount 10% mix	50 qt	30 qt	40 qt
Amount lemon juice from 10% mix	5 qt	3 qt	4 qt
Total amount new mix	70 qt	50 qt	60 qt
Total lemon juice in new mix	10 qt	8 qt	9 qt
Percent lemon juice in new mix	$\frac{10}{70}$ or 14%	$\frac{8}{50}$ or 16%	$\frac{9}{60}$ or 15%

✓ **Informal Assessment** Ask: If you want 3 qt of 15% lemon juice mix, how much would you mix of 10% and 25% mixes? (2 qt 10% and 1 qt 25%. That mixture yields $2 \times 0.1 + 1 \times 0.25$ or 0.45 qt lemon juice. $0.45 \div 3$ is 15%.)

2. If Stan's county had more sparrows last year than the next county, he is right to be concerned. If Stan's county had the same number or fewer sparrows last year than Sonya's county, Sonya is correct that the white-crown sparrow is okay. Suggest that students use **Guess and Check**.

Choose 100 as a sample number of sparrows in each county to examine the conditions. If Stan's county had 100 last year, and 50% fewer this year, that would be 50% of 100 or 50 fewer, so there are 50 birds this year. If the next county had 100 birds last year and increased 50%, they now have 150. In this scenario, in the two counties there were 200 birds last year and 200 birds this year.

Suppose that Stan's county had 110 birds last year and the next county had 90 birds. 50% of 110 is 55 fewer birds for a total of 55 birds in Stan's county. 50% of 90 birds is 45 birds for a total of 135 birds in the next county. In this case, there were 200 birds last year and 190 birds this year, a decrease of ten birds.

In the third example, Stan's county begins with fewer birds than the next county. Stan's county has 90 birds and the next county has 110 the first year. A 50% loss of birds in Stan's county leaves 45 birds. A 50% increase in birds in the next county yields 165 birds. There were 200 birds last year and 210 birds this year, an increase of ten birds.

4 USING PERCENTS

Challenge Your Mind

1. Dana is making lemonade for the school dance. She lost her recipe but remembers it is just lemon juice mixed with sugar syrup made from water and sugar. She makes one that is 25% lemon juice and one that is 10% lemon juice. She makes 20 qt of the 25% mixture and 60 qt of the 10% solution. She tastes each and decides that one is too strong and the other too weak. She thinks a 15% mixture would be ideal.

a. If she mixes the two together, what percent of the resulting 80 qt will be lemon juice?

b. How much of each mixture should she combine to get the most 15% lemon juice mixture possible?

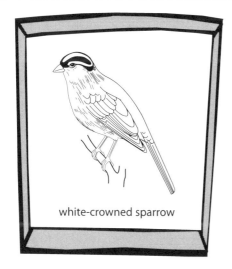

white-crowned sparrow

2. Stan is concerned because in the annual bird count, there were 50% fewer white-crowned sparrows counted in his county than in the previous year. Sonya says that is okay, because there were 50% more sparrows counted in the next county. Under what circumstances would Stan's concerns be justified? Under what circumstances would Sonya's conclusion be correct?

4 USING PERCENTS

Wrap It Up

Sunny Cities

Discuss the problem with the class before they begin to work on their own. Ask:

- What is the measure used in the problem to define the sunniest city? (The percentage of daylight hours the sun is covered by clouds.)

- What percentage of daylight hours are sunny in Yuma? (90%; 10% are cloudy, so 90% are sunny.)

Solutions

Part A Sacramento needs 527 more sunny hours and Key West needs 614 more sunny hours to surpass Yuma to become the sunniest US city. That would be a 15% increase for Sacramento and an 18% increase for Key West.

Part B Answers will vary. Answer should include New York City's data on at least one measure of weather and a local town's parallel measure. The answer should also include the percent of increase or decrease of the local town's measure to equal New York City's.

Possible Solution Paths

Part A Students may begin by finding the number of daylight hours per year by multiplying the average daylight hours per day by the number of days in a year. $12 \times 365 = 4,380$.

Then find the number of sunny hours for Yuma, Sacramento, and Key West by multiplying by the appropriate percent of sunshine. Yuma is 90% of 4,380 or 3,942. Sacramento is 78% of 4,380 or 3,416. Key West is 76% of 4,380 or 3,329.

Find out how many more hours of sunlight Yuma has than Sacramento and Key West by subtracting. Sacramento has $3,942 - 3,416$ or 526 fewer hours, so it needs 527 to surpass Yuma. Key West has $3,942 - 3,329$ or 613 fewer hours, so it needs 614 to surpass Yuma.

Finally to find the percent of increase needed, divide the needed hours by the current number of hours for each city. Sacramento needs a 15% increase ($527 \div 3,416 = 0.1543$). Key West needs an 18% increase ($614 \div 3,329 = 0.1844$).

Part B Make various reference materials available to your students as they research weather in New York City and a local town.

Assessment Rubric

3 The student accurately finds the number of daylight hours in a year, the number of sunny hours for each city, and the percent increase needed to surpass Yuma; the New York City and local town data are parallel and accurately compared; and the percent of change is accurate.

2 The student finds the number of daylight hours in a year, the number of sunny hours for each city, and the percent increase needed to surpass Yuma; the New York City and local town data are parallel; and the percent of change is presented. There may be a few arithmetic errors.

1 The student may find some, but not all of the needed data to find the percent increase, in some cases the wrong numbers are compared; the New York City and local town data are given, but are not complete or parallel; and the percent of change is inaccurate.

0 The student may have made some attempt at solving the problem but answers are either missing or incorrect; the New York City and local town data are missing and therefore student is unable to calculate percent of change.

4 | USING PERCENTS

Wrap It Up

Sunny Cities

Part A

The sunniest city in the continental U.S. is Yuma, Arizona. Yuma has the greatest annual percentage of possible sunshine. Clouds hide the sun only 10% of the time between sunrise and sunset 365 days a year. Sacramento, California has 78% of possible sunshine and Key West, Florida has 76% of possible sunshine.

Assume that the average amount of time between sunrise and sunset is 12 hours. How many more hours of sun per year do Sacramento and Key West need to have to surpass Yuma to take over first place? Round your answers to whole numbers of hours. What percents of change would that be for Sacramento and Key West?

Part B

Research the weather in your hometown or a large city nearby. Select data on annual rainfall, number of sunny days, amount of snow per year, average temperatures, or other weather information that you find. Compare that data to New York City's data on the same measures. How much would your city's data have to change to match New York City's? What percent of change would that be?

This section focuses on using fractions, decimals, and percents interchangeably and flexibly to solve problems. Students will have the opportunity to compare the ease of using one representation over another. They will discover that although they may prefer to work with one type of representation, in some problems a different representation may be a better choice, yielding simpler calculations.

Understanding Selecting Representation Problems

As students prepare to solve problems with fractions, decimals, and percents, they will need to consider what representation they are being asked for, what representation they prefer, what representation makes the most sense for the situation, or what representation makes the calculations easiest.

Students will use all four operations with fractions, decimals, and percents. They will also convert from one representation to another. As they solve problems that represent rational numbers in more than one way, students will need to consider which representation to use. Students should be encouraged to choose representations that allow them to calculate mentally whenever possible. Knowing key benchmark fractions, decimals, and percents will facilitate this process.

Calculators can be helpful in reducing the tedium of lengthy computations. If you allow students to use calculators for problems with heavy computation loads, they may see the patterns that emerge more readily since they can focus on those rather than on the procedure of doing the computation. However, many of the problems in this section are best done without a calculator as the problems focus on choosing the representation that streamlines the calculations. If students use a calculator, the impetus for simplifying the computation is greatly decreased. Choosing representations for their ease of use is a valuable skill in many situations.

Shoshi's science class watched a Webcast of an African watering hole. The students took shifts counting the animals over an eight-hour period. They counted 4 lions, 45 antelope, 23 warthogs, **and 18 waterbuck during that time. Express in simplest form the fractional part of the total number of animals that each kind of animal represents. Express the fractions as percents. Under what circumstances might you use each of these representations? Why?** To solve this problem, students will need to be able to:

- identify the part and whole in order to write the fractions.

- simplify the fractions.

- convert fractions to percents.

- use reasoning to determine when one representation may be preferable to another.

What is the area of a circle with a radius of 7 cm? What is the area of a circle with a diameter of 2.4 m? You may use 3.14 or $\frac{22}{7}$ for π. Which representation of π do you choose for calculating each area? Why? In a problem such as this, students must:

- know and be able to use the formula for the area of a circle and know how to find the radius, given the diameter.

- use what they know about operations to choose 3.14 or $\frac{22}{7}$ for π to make the ensuing calculations simpler.

- know how to multiply fractions and decimals.

- be able to express mathematical ideas.

As students work with selecting fractions, decimals, and percents, ask these key questions, as needed.

- How would you express that as a fraction? As a decimal? As a percent?

- If you chose a different representation would it make the calculations simpler?

- Does the problem have multiple steps?

- Is there more than one answer?

Solving Selecting Representation Problems

Your students will gain confidence and experience more success if they are selective about the specific representation they choose when solving a problem involving rational numbers. They will use logical reasoning and general problem-solving strategies in addition to skills involving fractions, decimals, and percents. Refer to *page vi* for a discussion of strategies.

Solving a Problem by *Making a Diagram* A diagram can organize the problem data making it easier to see what numbers to use in the calculations.

Example: The Singh family went on a 400-mile trip. The first day they drove $\frac{1}{4}$ of the way. The second day they drove 30% of the remaining distance. The third day they drove the rest of the way. How many miles did they drive each day?

Draw a Diagram to organize the information.

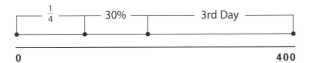

The first day the Singhs drove $\frac{1}{4} \times 400$ or 100 mi. The second day they drove 30% of the 300 remaining miles, 0.3×300 or 90 mi. Some students may prefer to use a fractional representation, $\frac{30}{100}$ for 30%, because it is easy to divide 300 by 100 and multiply that by 3. $\frac{30}{100}$ of 300 is 30×3 or 90. The third day they drove $400 - 100 - 90$ or 210 mi.

Assessment

☑️ **Informal Assessment** A suggestion for informal assessment will be found on each *Try It Out, Stretch Your Thinking,* and *Challenge Your Mind* page. The recommended question will help focus students' attention on one part of the problem-solving process.

Assessment Rubric An assessment rubric is provided for each *Wrap It Up*. Students' completed work may be added to their math portfolios.

Thinking About Selecting Representations

These problems require students to use their number sense and estimation skills in preparation for using percents to solve problems. Present one problem a day as a warm-up. You may choose to read the daily problem aloud, write it on the board, or create a transparency.

1. **What is 60% of $\frac{1}{5}$? What is $\frac{1}{5}$ of 60%? Explain why they are equivalent.**
   ($\frac{12}{100}$, 12%, answers will vary; Some students may draw a picture to show that they are the same. Others may say that these are both multiplication problems 60% $\times \frac{1}{5}$ and $\frac{1}{5} \times$ 60%, and since multiplication is commutative, the products must be the same.

   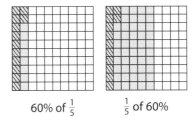

   60% of $\frac{1}{5}$ $\frac{1}{5}$ of 60%

2. **Fractions and percents are two different ways to name the same quantity. Name fraction and percent pairs that you know without figuring.**
   (Answers will vary. Commonly known pairs are $\frac{1}{2}$ and 50%, $\frac{1}{4}$ and 25% plus their multiples, $\frac{1}{3}$ and 33.3% plus their multiples, $\frac{1}{5}$ and 20% plus their multiples, $\frac{1}{10}$ and 10% plus their multiples, $\frac{1}{100}$ and 1% plus their multiples.)

3. **What is $\frac{2}{3}$ of 9? Of 2.1? What is 75% of 3.6?**
   (6, 1.4, 2.7; $\frac{1}{3}$ of 9 is 3, so $\frac{2}{3}$ is 2×3 or 6. Similarly $\frac{1}{3}$ of 2.1 is 0.7, so $\frac{2}{3}$ is 2×0.7 or 1.4. And 75% of 3.6 is the same as three-quarters of 3.6. $\frac{1}{4}$ of 3.6 is 0.9, so $\frac{3}{4}$ is 3×0.9 or 2.7.)

4. **What is 40% of 2.5? What is $\frac{2}{5}$ of 2.5? Which problem is easier for you to figure out? Why?**
   (1, 1, answers will vary; $2.5 \times 0.4 = 1.00$. $\frac{1}{5}$ of 2.5 is 0.5, so $\frac{2}{5}$ is 2×0.5 or 1. Students' answers to the last parts will depend on how they do the mental math.)

5. **What is 12.5% of 2? What is $\frac{1}{8}$ of 2? Which problem is easier for you to calculate mentally? Why?**
   (0.25, $\frac{1}{4}$ or 0.25, answers will vary; 2×0.125 is 0.25. Since 8 quarters makes $2.00, $\frac{1}{8}$ of 2 is 0.25.)

5 | SELECTING REPRESENTATIONS
Solving Problems with Selecting Representations

In this sections students will use fractions and percents to solve problems. They will also evaluate which representation makes the most sense to use.

Using the Four-Step Method

Find Out	
	○ How much of the music is downloaded expressed both as a fraction and percent. How long has Chinua been downloading music so far? Do you prefer fractions or percents to calculate the time that she has been downloading?
	● $\frac{30,556}{76,390}$ is not meaningful without estimating or expressing in simplest form. 00:01:54 is 1 min and 54 sec.
	○ $\frac{30,556}{76,390}$ is less than $\frac{1}{2}$ or 50%. So Chinua has been downloading less than 1 min and 54 sec.

Make a Plan	
	● Students know that 30,556K out of 76,390K are downloaded and that 1 min and 54 sec are left. They need to know what fraction or percent is downloaded to find out how long Chinua has been downloading.
	○ Students will choose a method to simplify the fraction. They may factor each number or use another method. To find the percent, they may do the division 30,556 ÷ 76,390 or simplify the fraction first and then divide to find the percent. They will **Use Logical Reasoning** to figure that if $\frac{2}{5}$ of the file is downloaded, the 1:54 represents $\frac{3}{5}$ of the time. There are several paths they may take using either the fraction or the percent.

Solve It	
	a. $\frac{2}{5}$; 40%; 30,556 ÷ 76,390 is $\frac{2 \times 2 \times 7,639}{2 \times 5 \times 7,639}$ or $\frac{2}{5}$. $\frac{2}{5}$ is 40%. Some students may divide 30,556 ÷ 76,390 to find the percent and then express it as a fraction. Accept all mathematically correct solution paths.
	b. 00:01:16; If the time so far is $\frac{2}{5}$, the time remaining is $\frac{3}{5}$. If 1:54 or 114 sec is $\frac{3}{5}$, $\frac{1}{5}$ is 38 sec. So $\frac{2}{5}$ is 76 sec or 00:01:16. Alternatively, if 40% represents the time so far, then 60% represents the remaining time of 114 seconds. If 60% of the whole time is 114, then 10% is 114 ÷ 6 or 19 seconds. Therefore 40% is 4 × 19 or 76 seconds. Accept all mathematically correct solution paths.
	c. Answers will vary.

Look Back	
	○ Have students compare their results with their estimates. Ask: Is your solution reasonable when compared with your estimate? How did estimating help you solve the problem?
	● A way to check the solution is to solve by a different method. Encourage students to share their solution methods in small groups.

5 | SELECTING REPRESENTATIONS

Solving Problems with Selecting Representations

Chinua is downloading music from the Internet. The download screen shows her progress. Assume that indicators are all accurate and that the download proceeds at a constant rate. (Yes, the time left is never right, but assume it is this time!)

a. Describe how much of the music is downloaded with a fraction in simplest form and with a percent.

b. Tell how much time Chinua has been downloading so far.

c. If you used fractions to answer part *b*, use percents to show a different solution path. If you used percents, try fractions. Which do you prefer? Explain.

Did you know that a byte is eight bits? A *bit* is the smallest information the computer can work with. A bit is the equivalent of a choice between two alternatives on or off/0 or 1. The word *bit* comes from *binary digit*. How many bytes is a kilobyte? a megabyte?

Find Out	○ What questions must you answer?
	• Is $\frac{30,556}{76,390}$ a meaningful fraction? How much time is 00:01:54?
	○ What are reasonable estimates?
Make a Plan	• What information do you have? What information do you need?
	○ How will you use the information to find the answer?
Solve It	• Find the answer. Use your plan. Keep a record of your work.
Look Back	○ How does your answer compare with your estimate?
	• How could you check if your answer is right?

5 | SELECTING REPRESENTATIONS

Try It Out

1. Make a Plan To help students plan their solution, ask:

- How can you make a circle graph? What information do you need? (Students may construct a circle graph using the fractions to partition the circle. Some students may use the percents and a protractor to make central angles for it. Students will multiply 360° by the percent to find the correct angle measure to represent each type of music.)

- Do you think your predictions will be valid? Why or why not? (Answers will vary. If the 750- and 200-student schools are similar to Cloud, the predictions could be close.)

Solution Path

a. $\frac{1}{5}$ rap, $\frac{3}{10}$ rock, $\frac{2}{25}$ country, $\frac{6}{25}$ pop, $\frac{4}{25}$ heavy metal, $\frac{1}{50}$ blues; There are $50 + 75 + 20 + 60 + 40 + 5$ or 250 students. $\frac{50}{250}$ or $\frac{1}{5}$ chose rap. $\frac{75}{250}$ or $\frac{3}{10}$ chose rock. $\frac{20}{250}$ or $\frac{2}{25}$ chose country. $\frac{60}{250}$ or $\frac{6}{25}$ chose pop. $\frac{40}{250}$ or $\frac{4}{25}$ chose heavy metal. $\frac{5}{250}$ or $\frac{1}{50}$ chose blues.

b.

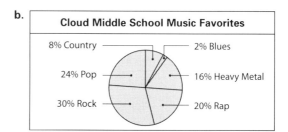

Cloud Middle School Music Favorites

8% Country — 2% Blues
24% Pop — 16% Heavy Metal
30% Rock — 20% Rap

c. 150 rap, 225 rock, 60 country, 180 pop, 120 heavy metal, 15 blues, answers will vary; Since the population of the school is 3 times Cloud's, it is easy to multiply by 3. Students may choose another solution path.

d. 40 rap, 60 rock, 16 country, 48 pop, 32 heavy metal, 4 blues; Since the population of the school is $\frac{4}{5}$ or 80% of Cloud's, find $\frac{4}{5}$ or 80% of each number. $\frac{4}{5} \times 50 = 40$, $\frac{4}{5} \times 75 = 60$, $\frac{4}{5} \times 20 = 16$, $\frac{4}{5} \times 60 = 48$, $\frac{4}{5} \times 40 = 32$, $\frac{4}{5} \times 5 = 4$. Some students will choose to find the percent instead of the fraction. Have students share and give reasons for their solution paths.

2. a. 0.12 is between $\frac{1}{16}$ and $\frac{1}{8}$ (closer to $\frac{1}{8}$), 0.6 is between $\frac{9}{16}$ and $\frac{5}{8}$; The most direct path is to convert the fractions to decimals. For example, $\frac{1}{16}$ is 0.0625 and $\frac{1}{8}$ is 0.125, so 0.12 is between $\frac{1}{16}$ and $\frac{1}{8}$. Similarly, $\frac{9}{16} = 0.5625$ and $\frac{5}{8} = 0.625$ so 0.6 is between $\frac{9}{16}$ and $\frac{5}{8}$.

b. $\frac{1}{4}$; Counting by eighths from $\frac{1}{8}$ to $\frac{5}{8}$ is 4 spaces. 25% is $\frac{1}{4}$ and $\frac{1}{4}$ of 4 spaces is 1 space. One space to the right of $\frac{1}{8}$ (counting by eighths) is $\frac{2}{8}$ or $\frac{1}{4}$.

c. Answers will vary. It is important for students to realize that although they may prefer one representation, there are times when a different one is more efficient.

3. a. 144; Since $33\frac{1}{3}\% = \frac{1}{3}$, then 48 represents $\frac{1}{3}$ of the students, so there are 48×3 or 144 students.

b. 7 students; $144 \times \frac{1}{4}$ is 36; $0.194 \times 36 = 7$.

c. 125 students, 180 students; $144 \div 1.15$ is 125. $144 \times \frac{5}{4}$ is 180. Students may use **Guess and Check** to find the solutions. For example they may guess 120, calculate 15% of it, add that to 120 to get 138 students (close but not 144) and continue this process.

✅ **Informal Assessment** Say: Seventy-two students went on a field trip to Atlanta. If 12 students got sick on the Atlanta trip, what fraction of the students who went on the trip got sick? What percent of the total enrollment at Journey got sick? ($\frac{1}{6}$, 8.3%; $\frac{12}{72} = \frac{1}{6}$ and $\frac{12}{144}$ is $0.08\overline{3}$.)

5 | SELECTING REPRESENTATIONS
Try It Out

1. The local radio station conducted a survey of students at Cloud Middle School. They were asked to select their favorite type of music from six choices. 50 students chose rap, 75 chose rock, 20 chose country, 60 chose pop, 40 chose heavy metal, and 5 chose blues.

a. What fraction of students chose each type of music?

b. Make a circle graph that shows the percentage of students that chose each type.

c. Using the data, predict how many students in a 750-student school would choose each type of music. Tell how you made your prediction and why you chose the method you used.

d. Answer the same questions as in **c**, but base your prediction on a 200-student school.

2.
a. Place 0.12 and 0.6 on the number line.

b. What fraction is 25% of the way between $\frac{1}{8}$ and $\frac{5}{8}$?

c. In this problem, when did you use fractions? Percents? Decimals? Why?

3. 48 or $33\frac{1}{3}$% of Journey Middle School students got on a bus to Atlanta.

a. How many students are there in the whole school?

b. One-fourth of Journey Middle School students do home study, and 19.4% of those home study students use a computer course. How many use the computer course?

c. This year's enrollment is an increase of 15% over last year. Next year, the enrollment is going to be $\frac{5}{4}$ of this year's enrollment. How many students were there last year? How many next year?

5 | SELECTING REPRESENTATIONS
Stretch Your Thinking

1. Find Out To help students understand the problem, ask:

- What is a radius? A diameter? (The line segment from the center of a circle to any point on the circle. A line segment that passes through the center of a circle and whose endpoints are points of the circle.)

- If you know the length of the diameter, how do you get the length of the radius? (divide by 2 since the radius is half the diameter)

Solution Paths

Answers will vary on which value of π is used for each circle. The fact that there is a choice

and that the ease of calculation is affected by the choice of the value of π is the focus of this problem.

a. 28.26 square units; The area is 3.14×3^2 or $3.14 \times 9 = 28.26$. Some students may choose to use $\frac{22}{7}$ for π.

b. 154 square units; The area is $\frac{22}{7} \times 7^2$ or $22 \times 7 = 154$. $\frac{22}{7}$ may be more efficient here because students can simplify the problem before they do the calculations.

c. 16.61 square units; The area is $3.14 \times (\frac{4.6}{2})^2$ or 3.14×2.3^2 or 16.61. Some students may use $\frac{22}{7}$ for π.

d. $38\frac{1}{2}$ square units; The area is $\frac{22}{7} \times (\frac{7}{2})^2$ or $11 \times \frac{7}{2}$ or $38\frac{1}{2}$.

---

2. The first store

The first store's price is $459 \times \frac{2}{3} \times 0.95$ or $290.70. The second store's price is $459 \times 0.7 \times 0.935$ or $300.42. The third store's price is $459 \times 0.8 \times \frac{5}{6}$ or $306. $290.70 is the lowest price.

You may want to discuss with students why $\frac{1}{3}$ off is the same as multiplying by $\frac{2}{3}$ or why 5% off is the same as multiplying by 95%. Use the distributive property to show that $P - \frac{1}{3}P$ (where P is price) = $P(1 - \frac{1}{3}) = \frac{2}{3}P$. The math is similar for 5% off.

3. 24%, 36% the second day

Some students may **Make a Picture or Diagram** to help them sort out the details of this problem.

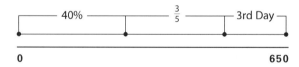

The first day the Twongs drove 650×0.4 or 260 mi, leaving $650 - 260$ or 390 mi. The second day they drove $390 \times \frac{3}{5}$ or 234 mi, leaving $390 - 234$ or 156 mi for the third day.

$\frac{156}{650} = 0.24$, $\frac{234}{650} = 0.36$.

---

4. Pieces totaling $\frac{3}{16}$, such as D or F, pieces totaling close to $\frac{21}{32}$, such as $D + E + F + H + I$, other combinations are possible

There are several solution paths. Students may name each piece as some number of 32nd and find that $\frac{1}{5}$ is between $\frac{6}{32}$ and $\frac{7}{32}$, closer to $\frac{6}{32}$ or $\frac{3}{16}$. Similarly $\frac{2}{3}$ is between $\frac{21}{32}$ and $\frac{22}{32}$, closer to $\frac{21}{32}$. Another way is to find the decimal equivalents of each of the fractional areas and the goal fractions, then find the closest combinations.

☑ **Informal Assessment** Ask: What combination of pieces would be close to $\frac{5}{6}$? ($A + B + C + E + F + G + H + I$ or all the pieces except for F. Since $\frac{1}{6}$ is close to $\frac{1}{5}$, all the pieces minus one that is close to $\frac{1}{5}$ will be close to $\frac{5}{6}$.)

5 | SELECTING REPRESENTATIONS

Stretch Your Thinking

1. Find the areas of the circles.

Remember that the formula for area of a circle is *Area* = π × the square of the radius or $A = \pi r^2$. Use 3.14 or $\frac{22}{7}$ for π.

Think about which approximation of π is easiest to use as you calculate each of the areas of the circles. In your answer, give the area of each circle, whether you used 3.14 or $\frac{22}{7}$ for π, and a reason for each choice.

a.

$r = 3$

b.
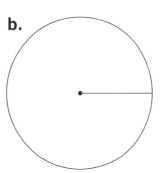
$r = 7$

c.

$d = 4.6$

d.
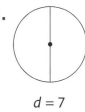
$d = 7$

---

2. Three different stores have the same $459 stereo on sale. One store advertises $\frac{1}{3}$ off and an additional 5% off if you buy it this week. The second store advertises 30% off, plus a rebate 6.5%. The third store says they will take 20% off, and you have a coupon for this store that says you can get any item for $\frac{1}{6}$ off the sale price. Where should you buy the stereo?

3. The Twong family took a 650 mi family driving trip to the beach. The first day they drove 40% of the way, the next day they drove $\frac{3}{5}$ of the remaining distance and the last day drove the rest of the way. What percent of the trip did they travel on the last day? What percent on the second day?

---

4. What piece or pieces of the rectangle would be close to $\frac{1}{5}$ of the total area? $\frac{2}{3}$?

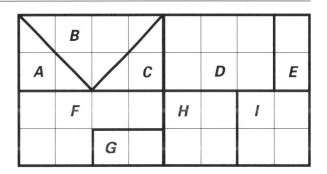

5 | SELECTING REPRESENTATIONS
Challenge Your Mind

1. **Find Out** To help students understand the problem, ask:

- What are vertices? (The corners of a polygon.)

- What are midpoints? (The point on a segment that divides it into two equal segments.)

- How much would half of the rectangle be worth? One fourth of the rectangle? (50¢; 25¢.)

Solution Paths

A is worth 25¢, B is worth 37.5¢, C is worth 12.5¢, and D is worth 25¢.

One way to solve the problem is to **Make a Picture or Diagram**. Divide the rectangle in half vertically and horizontally to make quarters.

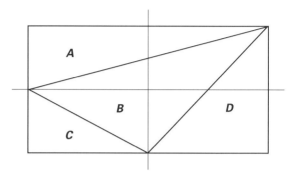

Once the rectangle is divided by the vertical and horizontal lines, students may see that A is $\frac{1}{2}$ of the top half of the rectangle, or $\frac{1}{4}$. Likewise, D is $\frac{1}{2}$ of the right half of the rectangle, or $\frac{1}{4}$. C is $\frac{1}{2}$ of the lower left quarter of the rectangle, or $\frac{1}{8}$. And B is the remainder.

Students may then express the fractions as their decimal equivalents to find the part of a dollar each section represents, or they may find the fraction of $1. $\frac{1}{4}$ is 25¢, $\frac{1}{8}$ is $12\frac{1}{2}$¢, so B is $1 − 25¢ − 25¢ − 12.5¢ or 37.5¢.

---

2. **263.45**

The first two clues tell the students that the number is greater than 600 × 0.4 or 240 and less than $\frac{2}{5}$ × 800 or 320. So the mystery number is between 240 and 320. That means that either 2 or 3 must be in hundreds place. Place the decimal point so that the first digit is the hundreds place.

The next clue tells students that the hundreds digits is $\frac{1}{3}$ of its tens digit. 3 is $\frac{1}{3}$ of 9, but 9 is not one of the choices. So the hundreds digit must be 2. If 2 is $\frac{1}{3}$ of the tens digit, the tens digit must be 6. So we have **2 6 ☐ . ☐ ☐**.

The fourth clue tells students that the tens digit is 150% of its tenths digit. 6 is 150% of 4. So the tenths digit is 4.

The fifth clue tells students that the tens digit is 100% greater than the ones digit. 100% greater means it is twice as much. 6 is twice as much as 3. Therefore, the ones digit is 3 and the hundredths digit must be the 5.

☑ **Informal Assessment** Ask: Suppose a number is made up of the digits 2, 4, and 6 and a decimal point. If the number is less than 10% of 300 but greater than $\frac{2}{3}$ of 9.5, what are the possible answers? (26.4, 24.6, 6.42; 10% of 300 is 30 and $\frac{2}{3}$ of 9.5 is 6.3. Therefore, the number must be between 6.3 and 30.)

5 | SELECTING REPRESENTATIONS

Challenge Your Mind

1. Suppose the vertices of triangle *B* connect the midpoints of two sides of the rectangle and the opposite corner of the rectangle. If the whole rectangle is worth $1.00, what is the value of each piece?

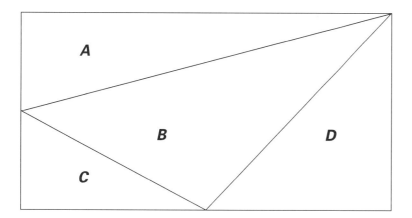

2. A mystery number uses the digits 2, 3, 4, 5, 6.

   It is more than 40% of 600.

   It is less than $\frac{2}{5}$ of 800.

   Its hundreds digit is $\frac{1}{3}$ of its tens digit.

   Its tens digit is 150% of its tenths digit.

   Its tens digit is 100% greater than its ones digit.

   What is the number?

5 SELECTING REPRESENTATIONS
Wrap It Up

News and Ads

Ask students to collect and bring in newspapers. If you begin a week in advance, only a few students will need to bring in papers in order for you to have enough for the entire class.

Discuss the problem with the class before they begin to work on their own. After you distribute newspapers, ask:

- Who can find an article or advertisement with a fraction or percent? Read it to the class. (Answers will vary.)

  Choose one of the articles or ads and rewrite it as described in the problem.

- Who can suggest a fraction or percent problem that could be written with the information in this article/ad? (Answers will vary)

Solutions
Answers will vary.

Assessment Rubric

3 The student accurately converts fractions and percents to the other representation; discusses how changing to the other representation affects the meaning; writes an appropriate fraction or percent problem based on their article or ad; and communicates the answer to the problem clearly and accurately.

2 The student converts fractions and percents to the other representation; discusses how changing to the other representation affects the meaning but may be vague; writes a fraction or percent problem based on their article or ad but it may not be very meaningful; and communicates the answer to the problem clearly with few calculation errors.

1 The student makes errors in converting fractions and percents to the other representation; does not communicate how the other representation affects meaning; and is unsuccessful at writing a problem based on the article or ad.

0 The student may find an ad or article, but cannot or does not complete the rest of the assignment.

5 | SELECTING REPRESENTATIONS
Wrap It Up

News and Ads

Find a news article and an advertisement that contain either percents or fractions.

Rewrite both the article and ad replacing percents with fractions or fractions with percents.

Does the article and ad still make sense? Which version do you like better, the original version or your revision? Why?

Use the information in the article or ad to make a word problem involving fractions and another using percents. Put the problem on one side of the paper and a detailed solution on the back.

6 USING RATIOS
Teaching Notes

This section focuses on using fractions, decimals, and percents to solve ratio problems.

Understanding Ratios

As students prepare to investigate the use of ratios, they should know how to find unit costs, rates (for example, mph), and simple probabilities. The concept of what two quantities are being compared is key to many of these problems. Most of the problems focus on fractions or percents, however decimals are found in problems involving measurement and money.

In science lab, a class of thirty students dissected owl pellets. Twenty-five students found mouse bones and five found bird bones. Assuming that is the usual ratio of bird bones to mouse bones, what is the probability that the next student will get bird bones in her owl pellet? Express the probability as a fraction and a percent.
To solve this problem, students will need to:

- understand that the probability is the number found over the total possible.

- express that fraction as a fraction in simplest form and as a percent.

Mark needs $2\frac{1}{2}$ lb of beans for his soup recipe. He finds fresh beans at 2 lb for $2.45 and canned beans at 12 oz for $0.89. Which is the most economical way for him to buy the beans? How much will he spend on beans?
In a problem such as this, students must:

- know how many ounces are in a pound and be able to convert from ounces to pounds or pounds to ounces.

- find a unit price, or prices for both choices at the same quantity, or quantities for both choices at the same price.

- multiply and/or divide fractions and decimals.

- compare amounts of money or quantities.

As students work with ratios, ask these key questions, as needed:

- What quantities are you comparing?

- Is the fraction the simplest form?

- Does the problem have multiple steps?

- Is there more than one answer?

Solving Ratio Problems

Your students will be more successful if they can apply logical reasoning and combine general problem-solving strategies with methods specific to ratio problems. Refer to *page vi* for a discussion of strategies.

Solving a Problem by *Making a Table* Tables are an appropriate way to list outcomes when students are working with probabilities.

Example: What combinations are possible with two number cubes if the first one has the numerals 1, 1, 1, 2, 2, 3, and the second one has the numerals 1, 2, 3, 4, 5, 6? Express the probabilities of rolling each combination as a decimal and percent.

Make a table that shows the numbers from the first die across the top and the second down the sides, with the combinations in the middle.

	1	1	1	2	2	3
1	2	2	2	3	3	4
2	3	3	3	4	4	5
3	4	4	4	5	5	6
4	5	5	5	6	6	7
5	6	6	6	7	7	8
6	7	7	7	8	8	9

Count the total number of combinations (36) and the number of times each combination occurs— three occurrences of 2, five of 3, six of 4, six of 5, six of 6, six of 7, three of 8, and one of 9. Then express the probability of each combination by comparing the number possible with the total combinations. The probability of rolling a 2 is $\frac{3}{36}$ or 8.3%, 3 is $\frac{5}{36}$ or 13.9%, 4, 5, 6 and 7 are each $\frac{6}{36}$ or 16.7%, 8 is $\frac{3}{36}$ or 8.3%, and 9 is $\frac{1}{36}$ or 2.8%.

Solving a Problem by *Using Logical Reasoning* and *Working Backward* Sometimes the most convenient way to solve a problem is to use the information given and work backward to get the answer.

Example: The tank of fish at the aquarium has a ratio of 4 guppies to 5 tetras. Two tetras were added to the tank, changing the ratio of guppies to tetras to 8:11. What is the total number of tetras and guppies after the addition of the tetras?

Using **Working Backward** and **Logical Reasoning** the final number of tetras must be a multiple of 11, (11, 22, 33, 44,…) Before the tetras were added, the number of tetras was a multiple of 5 (5, 10, 15, 20, 25, 30,…). The difference between the original number of tetras and the final number is 2, since 2 were added to the aquarium. The numbers 22 (multiple of 11) and 20 (multiple of 5) have a difference of two, so go back to the problem to see if those numbers work. Using the ratio of 4 to 5 and 20 tetras we can calculate that there are 16 guppies since 4:5 = 16:20. Adding 2 tetras gives 16 guppies and 22 tetras, and that is a ratio of 8:11. The total number is 16 + 22 or 38 fish.

Assessment

☑ **Informal Assessment** A suggestion for informal assessment will be found on each *Try It Out, Stretch Your Thinking,* and *Challenge Your Mind* page. The recommended question will help focus students' attention on one part of the problem-solving process.

Assessment Rubric An assessment rubric is provided for each *Wrap It Up.* Students' completed work may be added to their math portfolios.

Thinking About Ratios

These problems will help students use their number sense and estimation skills as preparation for using ratios to solve problems. Present one problem a day as a warm-up. You may choose to read the daily problem aloud, write it on the board, or create a transparency.

1. **To reconstitute frozen juice, you add three cans of water to one can of concentrate. What is the ratio of concentrate to water? How much water would you add to 2 cans of concentrate? To 0.25 cans of concentrate?**
   (1:3, 6 cans, $\frac{3}{4}$ can of water; Encourage students to share their ways of thinking about how much water to add.)

2. **If Omar lives 1.5 miles from school and it takes him 30 minutes to walk to school, how fast does he walk?**
   (3 mph; If he walks 1.5 miles in 30 minutes, he can walk 3 miles in 1 hour, so it is 3 mph. Some students may answer 1.5 mi per half hour. This is correct, but rates are usually expressed as something to one unit. In this case 3 mi:1 hr or 3 mph.)

3. **You have a bag with 20 cubes in it. 5 cubes are red, 5 are yellow, 5 are blue, and 5 are green. If you pull one cube out, what is the probability that you will pull a red cube?**
   ($\frac{1}{4}$ or 25% or 1 out of 4; There are 20 cubes and 5 are red, so you have a 5 out of 20 chance to get a red cube. Express the answer in the simplest form.)

4. **What is a better deal 3 for $1 or 2 for $0.69?**
   (3 for $1; 3 for $1 is $0.34 each—a better deal than $0.35 each. Retail prices are usually rounded up (so 0.333 is 0.34 not 0.33) if the purchaser buys less than the number for sale at the given price.)

5. **Which is faster, 4 pages in 5 minutes or 60 pages in an hour?**
   (60 pages per hour; 60 pages per hour is a page a minute. In 5 minutes that would be 5 pages, so it is faster than 4 pages in 5 minutes.)

6 | USING RATIOS

Solving Problems Using Ratios

This lesson will help students focus on using fractions and percents to express simple probability. Students reflect on whether fraction or percent representation is more meaningful to them in a specific example.

Using the Four-Step Method

Find Out	○ What is the probability Juan will pick a white marble? Express the probability as a fraction and percent. Is the fraction or percent more meaningful? ● 18 marbles are still in the bag. $28 - 10 = 18$. ○ Since there are fewer white marbles in the bag than red marbles, it is less likely that Juan will pick a white marble. The probability is less than $\frac{1}{2}$ or 50%.
Make a Plan	● Students know that there are 14 red and 14 white marbles to start. Three red marbles and 7 white marbles have been removed. They need to find out how many marbles are still in the bag and how many of those are white. ○ Add to find the number of marbles that were removed. Subtract that number from the total to find the number of marbles still in the bag. Subtract the white marbles removed from the beginning number of white marbles to find the number of white marbles remaining. Express the probability of picking a white marble by telling how many white marbles are in the bag compared to the total marbles in the bag. State the answer as a fraction and a percent.
Solve It	● $\frac{7}{18}$ or 38.9% chance of drawing a white marble, answers will vary; There were 28 marbles to start, and 10 were removed leaving 18 marbles. Originally 14 were white. Seven white were removed, leaving 7. So there is a $\frac{7}{18}$ chance of drawing a white marble. $\frac{7}{18}$ is 38.9%. Some students will prefer the $\frac{7}{18}$ expression because they can tell that it is a bit less than $\frac{1}{2}$, and once they know the numbers of marbles in the bag, the answer is done. Others may say that the 38.9% number is more meaningful because $\frac{7}{18}$ is not a familiar fraction.
Look Back	○ Have students compare their results with their estimates. Ask: Is your solution reasonable when compared to your estimate? ● One way to check that the answer makes sense is to have students **Act Out or Use Objects** to model the scenario of the problem. They could start with 11 red and 7 white cubes in a bag. The student will draw a cube from the bag and a partner can record the color. The cube should be replaced each time and this should be repeated 50 or 100 times. The number of white cubes picked is divided by the number of times they picked cubes from the bag. If they do this 100 times they will most likely draw a white cube between 35 and 45 times. The ratio should be close to 38.9%. This would be a good time to discuss why their answers may be different from the mathematical probability of 38.9%.

6 USING RATIOS

Solving Problems Using Ratios

Juan's class of 28 chooses teams by drawing red and white marbles out of a bag. They start with 14 red marbles and 14 white, and keep the marble they pick until the teams are final.

When it is Juan's turn to pick a marble, 3 of his classmates have already drawn red marbles, and 7 have drawn white marbles.

What is the probability that Juan will pick a white marble? Express the probability as a fraction and as a percent. Tell which expression of probability is more meaningful to you and why.

Find Out	○ What questions must you answer?
	● How many marbles are still in the bag?
	○ What are reasonable estimates?
Make a Plan	● What information do you have? What information do you need?
	○ How will you use the information to find the answer?
Solve It	● Find the answer. Use your plan. Keep a record of your work.
Look Back	○ How does your answer compare with your estimate?
	● How could you check to see if your answer is correct?

6 USING RATIOS

Try It Out

1. Make a Plan To help students plan their solution, ask:

- What does the problem ask you to find? (The most economical way to buy tomatoes.)

- How can the information in the problem help you find the solution? (Students may find the cost of 72 oz at each price, price per ounce of each, or prices for each at a multiple number of ounces, such as 12 oz or 36 oz.)

Solution Path

Two 36-oz cans is the most economical

Three 12-oz cans is 36 oz and they cost $0.90 × 3 − $0.03 or $2.70 − $0.03 or $2.67 for 36 oz. That is more than $2.59 for 36 oz.

---

2. Store-brand

$4.99 ÷ 3.75 = $1.33 per pound for the brand-name chicken. That is more than the store-brand chicken, which is $1.28 per pound.

3. 3-lb bag

$1.75 is less than 2 × $0.89 or $1.78.

---

4. Fresh corn

If one bag has three half-cup servings, then it contains $1\frac{1}{2}$ cups of corn. Since Alex needs 3 c of corn, he would need two bags, which will cost him $2.50. If he uses corn on the cob, he needs 6 cobs of corn at 3 for $1, that is $2.

5. 18 ribs, $2.58

From problem 1, students know that the increased recipe requires 72 oz of tomatoes, and since the original recipe called for 12 oz, Alex increased the recipe by 6. So he needs 6 times the original amount of celery. 3 × 6 is 18 ribs. To get 18 ribs of celery he must buy 2 bunches of celery. 2 × $1.29 is $2.58

---

6. $21.11

Tomatoes are 2 × $2.59 or $5.18. Chicken legs are 7.5 × $1.28 or $9.60. Celery is $2.58. Corn is $2.00. Onions are $1.75.

$5.18 + $9.60 + $2.58 + $2.00 + $1.75 = $21.11.

7. 2 c water, $\frac{1}{4}$ c rice, $\frac{3}{4}$ t thyme, $\frac{2}{3}$ c mushrooms, 12 oz tomatoes, $1\frac{1}{4}$ lb chicken legs, 3 ribs celery, $\frac{1}{2}$ c corn, $\frac{1}{3}$ c onion.

Divide each quantity for the increased recipe by 6 or multiply by $\frac{1}{6}$ to get the quantities for the original recipe.

✓ **Informal Assessment** Ask: If the original recipe serves 6 people, and Alex wanted to make exactly enough soup for 8 people, how much would he need of each ingredient? ($2\frac{2}{3}$ c water, $\frac{1}{3}$ c rice, 1 t thyme, $\frac{8}{9}$ c mushrooms, 16 oz tomatoes, $1\frac{2}{3}$ lb chicken, 4 ribs celery, $\frac{2}{3}$ c corn, $\frac{4}{9}$ c onion; He needs $\frac{8}{6}$ or $\frac{4}{3}$ of the recipe, so multiply each of the original ingredient quantities by $\frac{4}{3}$. Note: You may want to discuss with students the answers $\frac{8}{9}$ c and $\frac{4}{9}$ c. What would they actually use if they were making the recipe for 8? (probably 1 c of mushrooms and $\frac{1}{2}$ c of onions)

6 | USING RATIOS

Try It Out

1. Alex is going to make a big pot of soup for the school dinner. He increases the recipe his mom usually makes. The increased recipe requires 72 oz of canned tomatoes. Alex finds a store brand of tomatoes at $0.89 for 12 oz and 36 oz for $2.59. What is the most economical way to buy the tomatoes?

2. The soup also requires $7\frac{1}{2}$ lb of chicken legs. Alex finds store-brand chicken legs at $1.28 per lb. He finds packages of brand-name chicken legs weighing 3.75 lb each and marked $4.99 per package. Which is the more economical buy?

3. The increased recipe calls for 2 c chopped onion. Alex finds huge onions that cost $0.89 each. Each onion would make $1\frac{1}{2}$ c chopped onion. A 3-lb bag of small onions costs $1.75 and would yield at least 3 c chopped onion. Which would be the most economical way to buy the onions?

4. The increased recipe calls for 3 c of corn. Alex finds bags of frozen corn for $1.25. Each bag is labeled that it contains 3 half-cup servings. He finds fresh corn on the cob at 3 for $1. He knows that two cobs of corn will yield 1 c of corn kernels. What is the most economical way for him to buy the corn?

5. The original recipe calls for 3 ribs of celery. It also calls for 12 oz of canned tomatoes. How much celery does Alex need for the increased recipe? If a small bunch of celery has 12 ribs and costs $1.29, how much will Alex spend on celery?

6. Alex already has everything else he needs for the soup. How much does he spend at the store for soup?

7. Alex combines everything he bought at the store, plus 12 c water, $1\frac{1}{2}$ c rice, $4\frac{1}{2}$ t thyme, and 4 c mushrooms. What were the quantities of all the ingredients in the original recipe?

6 USING RATIOS

Stretch Your Thinking

1. **Find Out** To help students understand the problem, ask:

- What could the numerator of the fraction be? The denominator? (1, 2, 3, 4, 5, or 6; 3, 4, or 5)

- What is one fraction you could make? (Answers will vary, one possible answer is $\frac{1}{3}$.)

- What is a terminating decimal? What other kind of fraction is there? (A terminating decimal is one that can be expressed with a finite number of digits because after a certain point all the digits to the right become zeros. For example, $\frac{1}{8}$ is terminating because $1 \div 8$ is 0.12500000… and can be expressed as 0.125. Another kind of decimal is a repeating decimal. For example, $\frac{1}{3}$ is a repeating decimal expressed as $0.33\overline{3}$.)

Solution Paths

$\frac{7}{9}$ and 77.8%

Students may list all the combination of fractions and evaluate them, or they may use what they know about fractions to tell that $\frac{1}{3}, \frac{2}{3}, \frac{4}{3}$, and $\frac{5}{3}$ are the only fractions that are repeating. There are 6×3 or 18 fractions that can be made and $18 - 4$ or 14 of them are terminating, so there is a $\frac{14}{18}$ or $\frac{7}{9}$ or 77.8% chance that a fraction created by the methods given will be terminating.

---

2. **a.** 15 minutes

If $\frac{2}{5}$ of the way home takes 10 min, $\frac{1}{5}$ takes 5 min. He has $\frac{3}{5}$ of the way to go, so it will take him 3×5 or 15 min.

b. 3 mph

It takes Stefano 10 min + 15 min or 25 min to travel the 1.25 mi home. 1.25 in 25 min is 0.25 mi in 5 minutes. To find the number of miles per hour, multiply by 12. 0.25×12 is 3 mph.

3. **a.** 180 cm or 1.8 m

30×6 is 180 cm or 1.8 m.

b. 1.2 m

Students may find the height of the jump on Earth by multiplying by $\frac{1}{6}$, then the height on Mars by multiplying by 3, or they may figure that the jump on Mars is $\frac{1}{2}$ the jump on the Moon. $\frac{1}{6} \times 2.4 = 0.4$ and $0.4 \times 3 = 1.20$.

---

4. **a.** 10 times, 5 times; since $\frac{1}{2}$ of the spinner is 3, $\frac{1}{2}$ the spins will land on 3 and $\frac{1}{2} \times 20 = 10$.

b. Make sections $\frac{1}{8}, \frac{3}{8}$, and $\frac{4}{8}$ and label them 3, 1, and 2 respectively, $\frac{1}{2}$ or 50%;

Since 12.5% is $\frac{12.5}{100}$ or $\frac{25}{200} = \frac{1}{8}$, divide the spinner into eighths and label one of them 3. Similarly, 37.5% is $\frac{37.5}{100}$ or $\frac{75}{200} = \frac{3}{8}$ so three of the $\frac{1}{8}$ sections will be labeled as 1. Since $\frac{1}{8} + \frac{3}{8} = \frac{4}{8}$, then there are four $\frac{1}{8}$ sections remaining and will be labeled 2.

☑ **Informal Assessment** Ask: Make a spinner where out of 24 spins, 8 of the spins land on A, 12 of the spins land on B and 4 of the spins land on C. What is the probability (expressed as a percent) that a spin will land on B? (The spinner should show $\frac{1}{3}$ A, $\frac{1}{2}$ B and $\frac{1}{6}$ C. There is a 50% chance that any one spin will land on B.)

Name _____

6 USING RATIOS

Stretch Your Thinking

1. The numerator of a fraction is randomly picked from the set {1, 2, 3, 4, 5, 6} and the denominator of that fraction is randomly picked from the set {3, 4, 5}. What is the probability that the decimal representation of the fraction is a terminating decimal? Express the probability as a fraction and a percent.

2. Stefano makes it $\frac{2}{5}$ of the way home in 10 minutes.

a. How much longer will it take him to get home if he continues to travel at the same rate?

b. If Stefano's home is 1.25 miles from where he started, how fast did he travel?

3. Gravity is a force that pulls objects toward each other. The force of gravity between objects is related to the mass of the objects and the distance between them. The more mass an object has, the greater its gravitational force, and the less mass, the less its gravitational force.

a. The mass of the Moon is approximately $\frac{1}{80}$ of the mass of Earth and its gravitational force is $\frac{1}{6}$ of Earth's. If your vertical jump on Earth is 30 cm, what is it on the Moon?

b. Mars' gravitational force is $\frac{1}{3}$ that of the Earth. If your vertical jump on the moon is 2.4 m, what is it on Mars?

4. Each time you spin this spinner you can land on 1, 2, or 3.

a. If you used this spinner 20 times, how many times would you expect to land on 3? The 1?

b. How could you change the spinner so that the chance of landing on the 3 is 12.5% and the 1 is 37.5%? If the remaining area is labeled 2, what is the probability of landing on 2 expressed as a fraction and a percent?

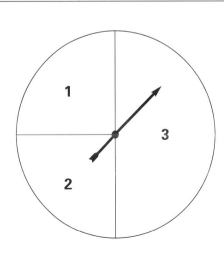

6 | USING RATIOS

Challenge Your Mind

1. Find Out To help students understand the problem, ask:

- What question does the problem ask you to answer? (What is the most impressive way to use percents or fractions to compare the download speed of the new MP3 player to the average MP3 player?)

- What information do you have? (The download times)

- What information do you need? (Different comparisons of the download times to see which one seems the most impressive

Solution Paths

Answers will vary.

Some students will convert 15 sec to 0.25 or $\frac{1}{4}$ min, and others will convert 15 min to 900 sec. Either way, the mathematics is similar.

- The new MP3 player takes only 1.7% of the time to download the same amount as the average MP3 player (15 ÷ 900 is 0.017 or 1.7%).

- Since 15 seconds is 885 seconds faster than 900 seconds, 885 ÷ 900 makes the new MP3 player 98.3% faster than the old.

- One album takes 0.25 minutes with the new player and that is 4 per minute or 60 per 15 min. This means that 60 albums can be downloaded in the same time as 1 with the old in a quarter hour.

- Finally, in one hour the average player downloads 4 albums compared to 240 albums per hour for the new player.

Different students will have different measures that impress them, they just need to justify why they think that measure is impressive and do the arithmetic correctly.

✓ **Informal Assessment** Ask: Which is faster, something that has a 150% increase in speed, or something that is 3 times faster? (3 times faster; 3 times faster means that if it was 100 mph before it is 3 times faster, or 300 mph. A 150% increase means that you find 150% of the initial speed and add it to the initial speed. 150% of 100 mph is 150 mph, and 150 + 100 is 250 mph. 300 mph is faster than 250 mph.)

2. 1,216 students; Students should realize that they are looking for a number that is a multiple of 77 and when 40 is subtracted from it, it becomes a multiple of 24. Multiples of 77 are 77, 154, 231, 308, 385, 462, 539, 616, 693, 770. Subtract 40 from each of these numbers and check to see if the difference is a multiple of 24. Students should realize that odd numbers can not be multiples of 24, so they can eliminate them. They should discover that 616 − 40 = 576, and 576 is a multiple of 24 (24 × 24 = 576). If 576 seventh graders signed up for the trip, and the ratio of seventh graders to eighth graders is 24 to 25 there are 576 × 25 ÷ 24 or 600 eighth graders. The total number of students going to Washington D.C. is 616 + 600 or 1,216 students. Tell students to double check to see that the first ratio 576:600 is the same as 24:25 (it is), and that when 40 is added to the number of seventh graders, the ratio 616:600 is the same as 77:75 (it is).

Problem Solving with Fractions, Decimals, and Percents   © Wright Group\McGraw Hill 0-7622-1254-3

6 USING RATIOS

Challenge Your Mind

1. The marketing manager for a new MP3 player knows that the average MP3 player takes 15 min to download an album. The new MP3 player can download the same album in 15 sec. The marketing person wants to report this information as the most impressive percent possible in an advertisement. Should the ad tell the percent of time it takes compared to the average player, how many times faster the new player is expressed as a percent, the number of albums per quarter hour or per hour, or another measure? Explain your thinking.

2. Of the middle school students who signed up for the school trip to Washington D.C., there is a ratio of 24 seventh grade students to 25 eighth grade students. Forty more seventh graders decide to sign up for the trip, changing the ratio to 77:75. How many students are going to Washington D.C. in all?

6 USING RATIOS

Wrap It Up

Game Designer

Discuss the problem with the class before they begin to work on their own. Ask:

- What does the problem ask you to do? (Create a game with at least four outcomes that have different chances of occurring. Express the probabilities for each outcome as a fraction and percent.)

- Discuss the example in the fifth bullet.

  What are the possible outcomes for the game? (4, 5, 6, 7, 8, 9, 10, 11)

  What is the probability of rolling a 9? ($\frac{1}{36}$ or 2.8%)

  What is the probability of rolling a 5? ($\frac{14}{36}$ or 38.9%)

  What other outcome has the same probability of occurring as 9? (11)

- Brainstorm with students. Write their list of ideas on the board. One example might be to use cubes. The cubes could be color-coded with dots so that the chance of rolling each color is different. Other examples are to make spinners or cards from which one can be drawn by the player with directions for a given action. If the cards are duplicated so that there are several of each kind, there will be different outcomes and thus different probabilities. There are many ways to inject different probabilities into the game.

Solutions

Game boards will vary

Assessment Rubric

3 The student creates a game board with outcomes that provide actions with at least four different probabilities; none of the four combinations of probabilities duplicates another; the game has a theme; and the student accurately expresses all probabilities as both fractions and percents.

2 The student creates a game board with outcomes that provide actions with at least four different probabilities; none of the 4 combinations of probabilities duplicates another; the game may have a theme; and the student expresses all probabilities as both fractions and percents, with a few arithmetic errors.

1 The student creates a game board with outcomes that provide actions with less than four different probabilities; at least one of the combinations of probabilities duplicates another; the game probably does not have a theme; and the student fails to expresses all probabilities as both fractions and percents, or if they are, there are many mathematical errors.

0 The student either does not create a game board or creates one but does not try to have different probable outcomes; probabilities, if any, are incorrect, showing little or no understanding of how to express probabilities as a fraction or percent.

6 | USING RATIOS
Wrap It Up

Game Designer

Design a board game.

Design Rules:

- Give your game a theme and a name.

- Decide how your game is to be played and what the rules are for it.

- Your game should have number cubes, spinners, or cards.

- There should be a minimum of four possible outcomes (example: a four-color spinner).

- The outcomes should have different probabilities for occurring. For example, if you use a pair of number cubes, one marked 1, 1, 1, 2, 2, 6 and the other marked 4, 4, 4, 4, 3, 5, the probability of outcomes is different.

- Your game should be original, creative, and ready to play.

- Tell the probability for each of the outcomes for one turn of play of your game. State the probabilities with both fractions and percents.

10 × 10 grid

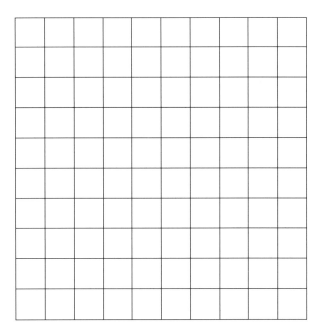

Conversion Charts

Length and Distance
1 km = 1,000 m = 100,000 cm = 1,000,000 mm
0.001 km = 1 m = 100 cm = 1,000 mm
0.01 m = 1 cm = 10 mm
0.001 m = 0.1 cm = 1 mm
1 mi = 1,760 yd = 5,280 ft = 63,360 in.
$\frac{1}{1,760}$ mi = 1 yd = 3 ft = 36 in.
$\frac{1}{3}$ yd = 1 ft = 12 in.
$\frac{1}{36}$ yd = $\frac{1}{12}$ ft = 1 in.

Capacity
8 fl oz = 1 cup (c)
2 c = 1 pint (pt)
2 pt = 1 quart (qt)
4 qt = 1 gallon (gal)

Mass and Weight
1 kg = 1,000 g = 1,000,000 mg
0.001 kg = 1 g = 1,000 mg
0.001 g = 1 mg
1 ton = 2,000 lb = 32,000 oz
1 lb = 16 oz